ELECTRONIC IGNITION
Installation, Performance Tuning, Modification

Ben Watson

This book is dedicated to Dennis Gearhart, Jim Moore, and Bill Hansen. These are the people who believed in me and put me in a position to tap on computer keys rather than twist wrenches in a shop that was either way too cold or way too hot.

First published in 1994 by Motorbooks International Publishers & Wholesalers, PO Box 2, 729 Prospect Avenue, Osceola, WI 54020 USA

© Ben Watson, 1994

All rights reserved. With the exception of quoting brief passages for the purposes of review no part of this publication may be reproduced without prior written permission from the Publisher

Motorbooks International is a certified trademark, registered with the United States Patent Office

The information in this book is true and complete to the best of our knowledge. All recommendations are made without any guarantee on the part of the author or Publisher, who also disclaim any liability incurred in connection with the use of this data or specific details

We recognize that some words, model names and designations, for example, mentioned herein are the property of the trademark holder. We use them for identification purposes only. This is not an official publication

Motorbooks International books are also available at discounts in bulk quantity for industrial or sales-promotional use. For details write to Special Sales Manager at the Publisher's address

Library of Congress Cataloging-in-Publication Data
   Watson, Ben.
   Electronic ignition/Ben
     Watson.
     p. cm. —(Motorbooks
     International powerpro series)
   Includes index.
   ISBN 0-87938-838-2
   1. Automobiles—Ignition—Electronic systems—Maintenance and repair. I. Title. II. Series.
TL213.W38   1994
629.25'49—dc20   93-6457

Printed and bound in the United States of America

On the front cover: An MSD Model 6AL high-performance ignition and an MSD Blaster 2 coil are shown with a selection of GM ignition components. Parts courtesy of Jeff Gual, MSD; Jim Kanan, Stillwater Motors; and Wynn Thacher, Chevrolet. *Eric Miller*

Contents

	Introduction	4
1	The Basics: Voltage and Its Sources	5
2	The Principles of Spark Ignition Combustion	21
3	The Point/Condenser Ignition System	30
4	AMC/Jeep: Breakerless Inductive Discharge (BID)	42
5	Solid State Ignition (SSI)	48
6	AMC/Renault Ducellier Electronic Ignition	53
7	Chrysler Passenger Cars and Light-Duty Trucks	57
8	Chrysler Hall Effects Ignition	63
9	Mitsubishi Electronic Ignition System	68
10	Electronic Lean Burn 1976-1977	73
11	Chrysler EFI and Optical Ignition	81
12	Ford Solid State Ignition (SSI)	95
13	Duraspark III	101
14	Thick Film Integrated (TFI) Ignition	105
15	General Motors Ignition System	118
16	Bosch Electronic Ignition System	144
17	Asian Electronic Ignition Systems	148
18	Aftermarket Ignitions	154
	Appendix	
	Ignition Specifications	158
	Index	160

Introduction

The automotive industry's first major move into the science of electronic engine controls was the electric ignition. The modern electronic ignition system differs little in its basic function from the original point/condenser system Charles Franklin Kettering designed in the early 1920s. However, when the electronic ignition systems were introduced in the early seventies, much mystery surrounded them. The automotive industry led us professional mechanics to believe that these systems were beyond the ability of mere mortals to troubleshoot. They led us to believe that if these ignition systems failed to function properly, it would take a team of M.I.T. physicists and half of the NASA technical staff to repair it. I exaggerate the facts, of course, but the emotional truth is not far from this thinking.

The reality of electronic ignition systems, and the reality of troubleshooting them, is that they are simple. I can remember losing customers in the mid-seventies because someone had told them that a mechanic needed $20,000-40,000 worth of specialized test equipment to maintain their new cars. In those days, and in these days, however, mechanics can diagnose electronic ignition systems accurately, efficiently, and professionally with the simplest of diagnostic equipment.

In writing this book, I realized that it will be the rare reader who will read beyond the section directly related to the car or cars he or she is working on. However, to get the most benefit from this book, I strongly recommend reading at least the chapters on electronics and point/condenser ignition before attempting to diagnose or repair your car.

The Basics: Voltage and Its Sources 1

Electrical systems have been part of the automobile since its inception at the end of the last century. First uses of electricity were the production of a high-voltage spark to ignite the air/fuel charge in the combustion chamber. As time went on, lighting systems, starting systems, radios, windshield wipers, and a wide variety of power accessories were added. In the late sixties, electronics began to appear on the passenger automobile. Electronic ignition systems, fuel injection systems, and emission and safety systems have become the standard of automotive technology.

As the sophistication of electrical and electronic systems has increased, the sophistication of the power sources for these components has had to increase as well. Back at the turn of the century when the use of electricity was limited to ignition, the magneto was the only electrical power source. When headlights were added, a source other than the high-voltage magneto was needed to power them. The storage battery was added.

With the addition of the storage battery, a whole new auto service business was born. The complexities of the lead-acid storage battery and its servicing were beyond the abilities of the typical mechanic of the day. Battery shops sprang up across the country. Since the storage battery needed to be recharged frequently to power the headlights, and eventually the starter, the generator was added. Again, the auto service industry was faced with new technologies.

From the early part of the century through the glory days of the automobile sports lover in the fifties and sixties, ignition systems changed little. Emission and fuel economy standards made the ignition system of the boat-tailed beauty, now gathering dust at the Honolulu airport, obsolete.

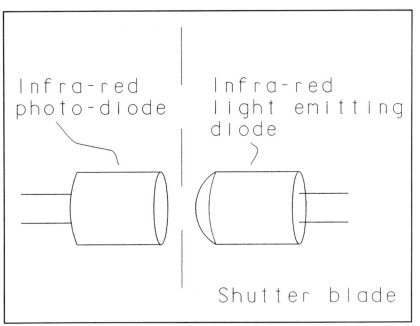

One of the latest electronic technologies that the automobile manufacturers have adopted is optical electronics. Although in use on aftermarket ignition systems since the mid-seventies, it was not until the mid-eighties that the manufacturers began to use it. A light emitting diode shines infrared light at a receiver, usually a photodiode. A disc with a series of holes interrupts the light beam. Rotating the disc creates a pulse. The frequency of the pulse is directly proportional to the rotational speed of the disc.

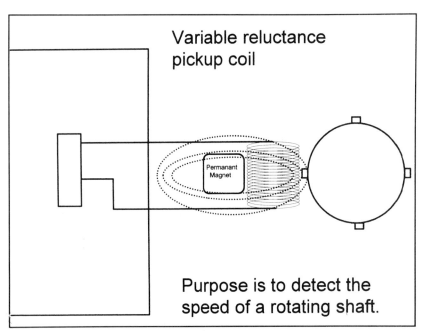

Since the late sixties AC pickup coils have been replacing the ignition points of old. There are other, more high-tech sensors than the AC pickup, but no other has undergone over thirty years of refinement. The AC pickup coil remains the most dependable. A magnetic field is cast across a coil of wire. A wheel with a series of protrusions, or teeth, is rotated through the magnetic field, causing the magnetic field to be distorted first toward the approaching tooth, then dragged along with the tooth as it moves away. This action creates an AC pulse.

Modern electronic engine control systems not only control the ignition timing but also constantly monitor and adjust the air/fuel ratio. This oxygen sensor produces a voltage that is directly proportional to the richness of the mixture. Contrary to what even many professional mechanics believe, it has no direct effect on the ignition system.

As more power equipment was added during the decades of the thirties, forties, and fifties, a more stable and dependable source of power was needed. The generator did an adequate job of keeping the storage battery charged, but it could not power the accessories at lower engine speeds. In the late fifties and early sixties, the alternator replaced the generator. In the more than thirty years since its introduction, the alternator has seen many improvements. In fact on many applications today, the alternator is computer controlled.

With all this increasing sophistication, there are still only three primary sources of voltage on today's automobiles: chemical, photoelectric, and electromagnetic.

Chemical Voltage Sources
The Battery

The automotive lead-acid storage battery has undergone many changes in the last few decades, but the principle of operation remains the same. Submerged in a solution of water and sulfuric acid are plates of two dissimilar metals, lead and lead peroxide. This condition sets up an imbalance of electron charges within the battery. The sulfate molecules try to force themselves into the negative lead plate while pulling oxygen atoms from the positive lead peroxide plate. The result is a surplus of electrons on the negative plate and a deficiency of electrons on the positive plate. The result is voltage.

The Oxygen Sensor

Today, there are two types of oxygen sensors used on automobile engines. The original, and by far the most common type, is the platinum oxygen sensor introduced in the late seventies on some Bosch fuel-injected engines. Today, most every car rolling off the assembly line has one.

Photoelectric Voltage Sources

There are rare instances where this new technology is used to generate a voltage. Light falling on some conductors excites the electrons in that conductor into motion, resulting in the generation. The first usage that may come to mind at this point is the solar cells used to power some experimental vehicles. Although this does hold some interesting possibilities for the future, today's uses are limited to only a few cases where the intensity of light needs to be measured.

Electromagnetic Voltage Sources

The most common method of creating electrical power or voltage on an automobile is through electromagnetism. Electromagnetism was at the heart of the magneto, generator, and alternator.

Electromagnetism has become even more important in recent decades with the advent of electronic control systems. Several of the sensing devices and most actuating devices operate on electromagnetic principles.

The four electromagnetic principles that play an important role in today's computer-controlled engine control systems are: induction, the electromagnet, solenoid action, and the Hall Effect.

Induction

The actual generation of voltage with electromagnetism is done through induction. When a conductor is passed through a magnetic field, electrons are pushed into motion. If the conductor is a straight piece of wire, electrons will be pulled from one end and pushed toward the other end, this creates the electron differential that is voltage. Connect the conductor to an electrical device and a current can flow.

Induction is used to produce voltages and currents in generators, alternators, and transformers such as the ignition coil. The AC pickup coil used in many electronic ignition systems also uses induction.

The pickup coil, or reluctance pickup, produces an AC voltage as a reluctor wheel rotates through a magnetic field. As the reluctor passes through, the magnetic field is distorted or warped across a coil of wire; this induces a voltage. Because the reluctor is rotating, the magnetic field is first distorted in one direction and then the other; this produces an AC signal.

The Electromagnet

Uses for electromagnetism do not end with the production of a voltage. When a current passes through a conductor, electrons have been set into motion. When electrons are set into motion, a magnetic field is generated around the conduit, or conductor through which they are passing. The strength of this magnetic field is directly proportional to the volume of electrons (size of the current flow) and the force pushing those electrons (voltage).

When the current carrying conductor is wound into a helix, the magnetic field generated by the conductor increases dramatically. Forming the helix around a soft iron core concentrates the magnetic lines of force and increases the strength of the field

One of the most common uses of an electromagnet, other than as an ignition coil, is in a relay. A relay uses a small current through the electromagnet to control a much larger current across the internal contact.

7

being generated. This is the basis for the electromagnetic relay.

A spring-loaded switch is located above an electromagnet. When the electromagnet is energized, the movable portion of the spring-loaded switch is attracted to the electromagnet and the switch changes position. When the movable contact changes position, the switch can be either opened or closed. If the switch is open when the electromagnet is de-energized, the relay is said to be normally open. If the switch is closed, then the switch is said to be normally closed.

Electromagnets are also fundamental to electric motors. There are two types of electric motors used on today's automobiles: the DC motor and the stepper motor.

The DC motor consists of a rotating armature which holds the conductor. As a current passes through the conductor, an electromagnetic field is generated which opposes the field of one or more permanent field magnets, resulting in rotation. The speed of rotation depends on the strength of the magnetic field. The strength of the magnetic field is directly related to voltage and current flow. DC motors are used effectively where either short- or long-term steady state rotational speed is required.

The design of the DC motor holds one inherent flaw. The current is carried to the electromagnetic windings through a set of brushes that ride on commutators on the armature. As the armature rotates, the brushes are slowly worn down thereby limiting the service life of the motor.

Applications for this type of motor include starter, power window, windshield wiper, and some idle speed control motors.

The primary difference in the construction of the stepper motor and the DC motor is the location of the windings and the

The DC motor has a wide variety of uses in a modern automobile; from the mundane starter to idle speed controllers and even in the fuel pump.

The stepper motor was a new type of motor that came into use during the eighties. It has a permanent magnet armature and an electromagnet field, the reverse of a conventional motor. In the stepper motor configuration, there are no brushes, which improves dependability. To the automotive engineer, its primary advantage occurs when the field coils are energized and the armature rotates a predictable amount, making precision control possible.

Solenoids have been around since the early fifties to control the starter. Today, solenoids are controlling everything from EGR operation to fuel flow to idle speed. Ford uses a solenoid on most of its fuel-injected engines to control idle speed.

permanent magnets. The stepper motor uses a stationary electromagnet while the armature holds the permanent magnets. This eliminates the need for brushes thereby increasing the service life of the component.

The primary disadvantage of this stationary winding motor is that to maintain a constant rotational speed, the current must pass through at least two sets of field windings. Furthermore, these field windings must be switched on and off at an extremely high frequency. Therefore, accurate and consistent motor speed requires electronic or computer control.

This disadvantage can be turned into an advantage, however. If two opposing sets of stationary field windings are used, each time a set is powered, the armature will rotate one-half revolution. When it reaches a point of magnetic neutrality it will stop. Therefore, precise positioning of controlled devices can be achieved.

The most common example of the use of a stepper motor is in the idle speed control motors used on late-model fuel-injected General Motors and Chrysler products. Here, the stepper motor controls a plunger position with a "worm gear." As a current is passed through one set of field windings, the armature will rotate one-half turn and stop. As the armature rotates, the plunger is moved in one step. Reversing the current flow through the field causes the armature to rotate in the opposite direction and the plunger moves out one step.

An important advantage of this type of motor is that it has no brushes to wear.

Solenoid Action

Although the solenoid and relay often perform similar tasks, the two are different types of devices. The solenoid, like the relay, incorporates an electromagnetic coil with an iron core to concentrate the magnetic field. The iron core, however, is spring loaded away from the center of the magnetic field. When the solenoid windings are energized, the iron core is pulled against spring tension into the center of the magnetic field. To perform a task, one end of the iron core is attached to a switch or a valve. When the solenoid is energized, the valve or switch either opens or closes.

There are three examples of

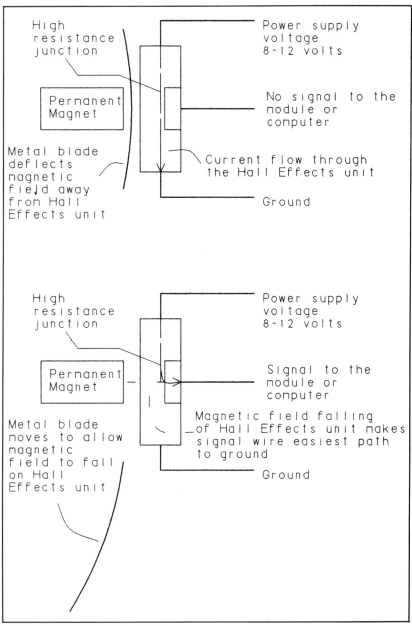

The Hall Effects sensor was first used in ignition systems in the early eighties. It uses a semiconductor device designed to be peculiarly sensitive to a magnetic field. A permanent magnet is situated so that it casts its field on the semiconductor. A set of vanes are connected to the distributor shaft, or any rotating shaft. As the vanes rotate, the magnetic field falls intermittently on the semiconductor. The semiconductor creates a voltage pulse, the frequency of which is directly proportional to the rotational speed of the shaft.

solenoid usage. First, it is used as a switch to control high current flow to the starter. Second, it is used as a valve to control vacuum to the EGR valve. Third, it is used as a valve to fuel flow into the intake manifold (an injector).

The Hall Effect

The Hall Effect, stated in technical terms: When a current is passed through a conductor and the conductor is placed in a magnetic field perpendicular to the direction of that current flow, then a voltage is produced which is perpendicular to the direction of that current flow.

Stated in less technical terms: When a magnetic field is placed perpendicular to a conductor with a current passing through it, the magnetic field either sucks voltage toward or pushes voltage away from the magnetic field.

This characteristic allows the Hall Effect to be an ideal method of detecting rotational speed and position. It has a significant advantage over the reluctance pickup coil in this since there is no minimum speed of rotation that can be measured.

Hall Effects devices are used to measure distributor shaft speed and position as well as that of the crankshaft and camshaft on many applications.

Measuring Volts
Units of Measurement

The basic unit of voltage measurement is the volt. Until the introduction of electronics onto the automobile, this single unit of measure, measured to the nearest tenth, was adequate. Many electronic devices produce very small voltages. In some devices, a change of only one-tenth of a volt or so can produce dramatic changes in their operating conditions. For this reason, measurement of automotive electronics is done in millivolts, units of one-thousandth of a volt.

Analog Meters

The analog voltmeter was for a long time the standard of not only the automotive industry but also the electronics industry. The analog meter derives its name from the way in which it displays information to the user. A needle sweeps across the numerical display to indicate voltage readings.

The primary advantage of the analog meter is its ability to show low frequency fluctuations in voltage by pulsations in the needle. Therefore, the analog voltmeter is an effective tool in tracking down intermittent circuit problems.

The primary disadvantage is that unless a technician uses an analog meter frequently they can be difficult to read accurately.

Digital Meters

The digital voltmeter displays its reading in a two-and-one-half to four digit display. This display is normally either LCD or LED.

The primary advantage of the digital display meter is that it is easy even for the infrequent user to obtain accurate readings.

The primary disadvantage is that instead of being a continuous readout, as with the analog meter, the digital voltmeter displays only sample readings. How frequently the meter samples the voltage being measured determines the meter's ability to find intermittent fluctuations in voltage.

High Impedance vs. Low Impedance

Impedance relates to the amount of resistance within the meter between the red lead and the black lead. A voltmeter is

In the early eighties the tool truck guys started pulling up in front of auto repair shops with a mission: to replace all analog meters with digital ones. While there are many uses for which the analog meter is inadequate, when it comes to measuring changing voltages, it is the perfect choice.

Some circuits carry currents that are so small many analog meters will put a drain on the circuit and distort the readings. For circuits carrying small currents, the digital meter is the perfect choice. This small, inexpensive meter will work as well for most circuits as a more expensive meter.

typically connected in parallel to the circuit being tested. A low impedance meter has a tendency to pull power away from the circuit being tested. This results in voltage readings lower than fact, particularly in circuits such as the oxygen sensor where the current flow is tiny. A high impedance meter pulls little power away from the circuit being tested. It is extremely accurate when used on low current flow circuits.

Voltmeter impedance is measured in volts per ohm. If the voltmeter is a 2k/voltmeter, and the meter is set on the 15 volt scale, the resistance between the red lead and the black lead will be 30,000 ohms. The typical high impedance meter will have an internal resistance of 10,000,000 ohms per volt. If the meter was placed on the 2 volt scale to read oxygen sensor output, the resistance between the red lead and the black lead would be 20,000,000 ohms.

Following is an example of the effect of voltmeter impedance on a circuit being measured: A 5 volt power supply is feeding voltage to two devices connected in series. One device has a resistance of 15,000 ohms. The second device has a resistance of 30,000 ohms. The voltmeter is set on the 15 volt scale, which gives it an impedance of 30,000 ohms (2,000 ohms per volt). Connecting the voltmeter on the input side of the second resistance to ground should yield a reading of about 3.3 volts. However, the voltmeter forms a parallel circuit with the 30k resistance. This newly formed parallel circuit lowers the resistance between the point being measured and ground causing the voltage at the point where the red lead is connected to also drop. The result is 2.5 volts. This is an error of nearly 25 percent.

To test most automotive electronic circuits, the meter should have a minimum of 50,000 ohms per volt. The issue of analog versus digital is purely a matter of choice and the need to see intermittent fluctuations providing the impedance requirements are met.

AC Voltmeter

The AC voltmeter has limited application to automotive troubleshooting. The one effective use is to test the reluctance type pickup. The traditional method of testing these devices is with an ohmmeter. The problem with this method is, although it tests the pickup coil for continuity, it does not test it for actual operation.

Connecting an AC voltmeter to a pickup coil and cranking the engine should produce an AC voltage of around seven-tenths of a volt.

The Oscilloscope

Long the exclusive domain of the electronics technician, the oscilloscope has begun to make itself at home in the world of auto repair and service. The oscilloscope is a voltmeter that displays changes in voltage over time. The vertical axis of the scope indicates voltage, the horizontal axis indicates time.

Low-impedance meters, such as the analog meter on the right, add load to the circuit they are testing. If the current flowing through the circuit being tested is small, the low-impedance meter can actually ground out the circuit and read zero volts. High-impedance meters, such as the one on the right, steal little power from the circuit being tested. As a result, the readings are not distorted and are accurate.

Although still not a common piece of equipment in most auto repair shops, the low-voltage oscilloscope has many uses in troubleshooting electronic ignition systems. The oscilloscope provides accurate measurement of voltage that changes over time making it the perfect tool for checking output signals from a pickup coil or Hall Effects.

Amps

If the pressure pushing electricity through a conductor is known as voltage, then the volume at which the electricity is flowing is called amps. A common analogy used to explain electrical circuits is city water pressure. Volts are like pounds per square inch; amps are like gallons per minute.

The measurement and control of amps has always been important in automotive electrical systems. The principle tests of battery performance are the current capacity tests. One of the principle tests of starter performance is the amperage draw test. In both of these cases, the volume of current being measured is extremely large. The automotive service industry has traditionally dealt with currents of 1 to 500 amps and even more.

The advent of automotive electronics has brought currents

much smaller than the auto technician has ever dealt with before. The oxygen sensor circuit at times carries a current of less than .0000000005 amps (less than one half-trillionth of an amp!). Although currents in these ranges behave exactly the same as larger currents, their perceived behavior can be different.

Normally, when a conductor carrying a current accidentally touches ground prematurely the result is dramatic. Since the power supply for most circuits on the automobile is either the battery or the alternator, a conductor shorting to ground can easily be overloaded. When extreme heat builds up in an overloading circuit, the conductor can actually melt. Most of the electronic circuits on the automobile have power supplies with a maximum potential far less than what the conductor can carry. When a low current circuit becomes grounded the conductor is not affected. The effect on the power supply, however, can be devastating. The power supply could be overloaded to the point of destruction.

Measuring Current Flow
Units of Measurement
The basic unit of measure for current flow is the amp. To the engineer, 1 amp of current flow means 6.28 billion billion electrons per second passing any point in the circuit when pushed by 1 volt. In servicing automotive electronic systems, there are few times when amperage will need to be measured directly.

Units of measurement for current flow are amps and milliamps. The milliamp is one-thousandth of an amp.

The Inductive Ammeter
In measuring current flow, there are two types of ammeters: the inductive and the shunt. The inductive type is usually chosen for large current flow applications. A clamp surrounds the conductor being measured. The clamp measures the strength of the magnetic field being generated by the current passing through the conductor. The strength of the magnetic field is then converted into amps. The advantage of the inductive ammeter is its ease of use. The conductor in which the current is being measured does not have to be disconnected. The disadvantage of the inductive ammeter is that the strength of the magnetic field in low current circuits is too weak to measure accurately. Below 10 amps, the inductive ammeter is a poor choice.

The Shunt Ammeter
The shunt ammeter must be connected in series with the current being measured. This requires that the conductor be disconnected. This makes the shunt ammeter more difficult to use. The advantage of the shunt ammeter is that all the current passing through the circuit being tested passes through the meter. It is extremely accurate down to milliamps.

When measuring amps in a large current flow, the inductive ammeter is the ideal choice. It requires no special connections. When measuring very small currents like those found in automotive computer circuits, the proper choice is a shunt type milliamp meter.

Resistance
Resistance is anything that impedes or inhibits the flow of electricity through a conductor. All conductors have some natural resistance in them. Resistance can be demonstrated with a simple high school physics experiment.

Using a number 2 pencil, draw a dark line on a piece of standard notebook paper. Place the black lead of an ohmmeter at one end of the line. Place the red lead of the ohmmeter immediately next to the black but do not allow the leads to touch. Note the ohmmeter. Now slowly move the red lead along the pencil line away from the black. Note that the ohmmeter reading increases. This is resistance.

Diameter, Length, Temperature, Material
Diameter, length, temperature, and material determine exactly how much resistance a given conductor will have.

The larger the diameter of a conductor, the less resistance there will be. The less resistance there is in a conductor, the greater current it can carry. It is standard practice to use small wires of low current circuits such as sensor circuits and larger wires of higher current flow circuits such as actuator circuits. Compare this to a water hose. A 0.5in hose cannot carry as large a flow, or current, as a 4in hose.

The length of the conductor will affect its resistance. Resistance increases with the length of the conductor in the same way that the ohmmeter reading increases as the red lead is moved along the pencil mark. The basic wire resistor, such as those used for ballast resistors, uses this principle.

For most conductors, as its temperature increases so does its resistance. One notable exception to this is battery electrolyte.

A peculiar phenomenon occurs as a current passes through a conductor. The current generates heat. As the temperature of the conductor increases, the resistance of that conductor also increases. The increasing resistance tends to impede current flow. If the temperature of the conductor rises beyond the melting point for the material forming the conductor, then the conductor will burn open.

Note: When conductors are cooled to extremely low temperatures, such as -200deg C or less, they become superconductors. This will permit extremely large currents to pass through small, lightweight conductors. Although there are currently no automotive applications for this technology, it does hold promise for emission-free electric automobiles in the future.

The material that forms the conductor may be the most critical in determining its resistance. The best conductor is silver. One of the worst of the commonly used conductors is aluminum. Gold is used as a conductor on many critical circuits or connections. Although other materials have conductive properties superior to gold, gold does not corrode. Corrosion at connection points increases resistance in a circuit. Because of extremely good conductive properties and a relatively high resistance to corrosion, copper is the most frequently used conductor in automotive circuits.

Resistors

Opening a device with electronic circuitry will reveal that the most common electronic component is the resistor. Resistors perform many functions in a circuit. For automotive purposes, two of its functions are current limiting and creating a voltage drop.

Any resistance in a circuit reduces current flow in that circuit. Resistors are placed in many automotive electronic circuits to limit current flow. When a wire becomes grounded, resistance in the series circuit is reduced. If reduced enough, the increased current flow from the power supply could either damage the power supply or the current carrying wire. Putting a current limiting resistor in the series circuit inside the power supply module will prevent this.

When a resistance is placed in a circuit carrying a current, voltage will be reduced as the current passes through the resistance. By using resistors that change value as events occur, it becomes possible for a computer to monitor these events through the changes in voltage.

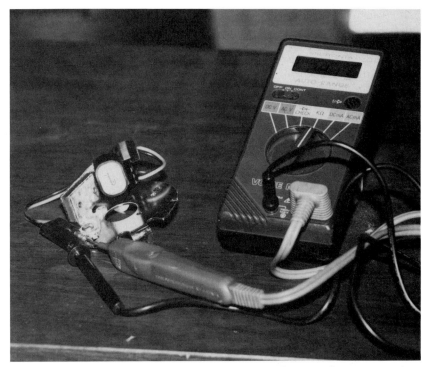

The ohmmeter is used to check electrical resistance. Here an ohmmeter is being used to test the resistance in a Chrysler pickup coil.

Measuring Resistance

Resistance is measured in ohms. One ohm is the amount of resistance it takes to limit 1 volt applied to a conductor at 1 amp of current flow.

An ohmmeter is used to measure resistance. It contains an internal voltage source, such as a 1.5 volt battery, and is connected in series with the component being tested. All power supplies must be removed from the component being tested. The ohmmeter passes a known current through the component, measures the outbound voltage, measures the return voltage, and uses the voltage drop to calculate the resistance.

Be careful when testing components that have a polarity, such as diodes. Some ohmmeters apply the positive voltage to the black lead when testing while others apply it to the red lead. Familiarity with the meter being used will prevent condemning a good component.

Watts

Watts is a measurement of electrical power. One horsepower is equal to 746 watts. There are watt ratings for power consumption for many electrical and electronic components. Resistors have watt ratings to indicate the amount of power they will dissipate; motors and lights have watt rating to indicate how much power they will consume. Beyond this, there are few times when watts will come up in troubleshooting or diagnostics.

Ohm's Law

Should the reader ever find himself speaking before a group of auto mechanics, and should the reader desire to elicit a group groan, the reader should say that he is going to discuss ohms law. We are now going to discuss ohms law.

There is a mathematical relationship between volts, amps, and ohms. Stated in the simplest terms, it takes 1 volt of electrical pressure to push 1 amp of current through 1 ohm of resistance. For the automotive technician, there are two important facts concerning ohms law. First, assuming we keep the voltage the same, as the resistance in a circuit decreases the current flow in the circuit increases. Second, assuming we keep the voltage the same, as the resistance in a circuit increases the current flow in the circuit decreases.

The second fact of ohms law is what enables the current limiting resistors used in computer circuits to prevent overload of the power supplies and voltage regulators.

Kirchoff's Second Law

There is another electrical law built on the principles of ohms law. Kirchoff's second law (we do not need to address his first) states "The algebraic sum of the voltage drops in a series circuit equals source voltage."

This means that as a current passes through resistances in a series circuit, voltage will be lost. The amount of voltage lost is directly proportional to the resistance. If a circuit is powered by 5 volts, and if there are five equal resistances in the circuit, there will be 1 volt lost as the current passes through each resistance.

A slightly more scientific explanation involves the notion of voltage drop per ohm. If the source voltage in a circuit is 5 volts, and the total resistance in that circuit is 5,000 ohms, the drop in voltage as the current passes through each ohm would be 1 millivolt (0.001 volt). Therefore, if the 5,000 ohms consists of a 2,000 ohm resistor, a 1,000 ohm resistor, and a second 2,000 ohm resistor, then the voltage drop across each resistor will be as follows:

2,000 ohms x 0.001 volts = 2 volts drop
1,000 ohms x 0.001 volts = 1 volt drop
2,000 ohms x 0.001 volts = 2 volts drop

The sum of the three voltage drops in the above example equals 5 volts, or source voltage. Implied in this is that if the voltage is measured after the first resistance the result would be 3 volts. Measuring after the second resistance would yield 2 volts and after the third resistance, zero.

This law remains in effect regardless of the number or size of resistances in a series circuit. To take it to an extreme, suppose there is a series circuit with ten resistances. The first nine are 1 ohm resistances while the last is a 10 megohm (10,000,000 ohm) resistance. The source voltage is 12 volts.

The total resistance of the circuit is:
10,000,000+1+1+1+1+1+1+1+1+1 or 10,000,009 ohms.

As the current passes through each ohm, the voltage drop is 12 divided by 10,000,009 or 0.00000119999892 volt.
1 ohm x 0.00000119999892 volt = 0.0000012 volt drop leaving 11.9999988
1 ohm x 0.00000119999892 volt = 0.0000012 volt drop leaving 11.9999976
1 ohm x 0.00000119999892 volt = 0.0000012 volt drop leaving 11.9999964
1 ohm x 0.00000119999892 volt = 0.0000012 volt drop leaving 11.9999952
1 ohm x 0.00000119999892 volt = 0.0000012 volt drop leaving 11.9999940
1 ohm x 0.00000119999892 volt = 0.0000012 volt drop leaving 11.9999938
1 ohm x 0.00000119999892 volt = 0.0000012 volt drop leaving 11.9999926
1 ohm x 0.00000119999892 volt = 0.0000012 volt drop leaving 11.9999914
1 ohm x 0.00000119999892 volt = 0.0000012 volt drop leaving 11.9999902
10,000,000 x 0.00000119999892 = 11.9999892 volts drop leaving 0 volts

---

Although the decimal places limit the accuracy of the mathematics, it is clear from the above that a proportional loss of voltage occurs as the current passes through the series circuit. After the last resistance, at ground, the voltage is always zero.

The Sine Wave

A sine wave is produced when a voltage gradually increases over time, reaches a peak, then gradually decreases. The classic sine wave resembles a series of evenly spaced rolling hills and valleys. However, this classic pattern is seldom seen when looking at waveforms on an oscilloscope. More often, the waveform looks like a jagged mountain range.

A common misconception about a sine wave is that it occurs only in an AC circuit. Although this is true in the automotive electronic systems where sine waves are tested, it may not always be true.

Production of a Sine Wave

There are two major components in automotive systems that produce sine waves. The first and the oldest is the alternator. The second, the stator, produces an AC sine wave that is converted into a DC voltage by the rectifier bridge. Although this is a major device, it is not within the scope of this book.

One common sensing device is the variable reluctance transducer, commonly known as a pickup coil. The sine wave produced by a pickup coil can be viewed on an oscilloscope but is more easily tested with an AC voltmeter.

Other than checking for the presence of the signal, sine waves are not normally tested in automotive electronics.

The Square Wave

Viewed on an oscilloscope, a square wave appears as a series of right angles. Like the sine wave, the square wave may appear broken or erratic but will follow the basic right angle pattern.

Electronic devices that produce a square wave include Hall Effects devices, optical position sensors, and "pulse generators" such as the GM MAF sensor. Unlike the sine wave, the square wave is frequently measured when troubleshooting automotive electronics. Four characteristics of the square wave that are commonly measured are amplitude, frequency, duty cycle, and pulse width.

Amplitude

Of the four measurable characteristics of a square wave, this is the one measured the least frequently. Amplitude is the amount of change in voltage from the lowest part of the square wave to the highest part. If the highest part of the square wave was 5 volts, the amplitude is 5 volts. This, of course, assumes that the lowest part of the square wave is at 0 volt. If the lowest part of the square wave was at 2 volts, then the amplitude would be 3 volts.

Since there is no way of determining the lowest point of the square wave with a voltmeter, the only accurate way of measuring amplitude is with an oscilloscope. With the waveform displayed on the scope, observe the voltage of the lower horizontal line and subtract from the voltage of the higher horizontal line.

Effects of Changes in Amplitude on Average Voltage

When the amplitude increases, the average voltage will also increase. When a voltmeter is connected to a circuit where the amplitude is changing, the voltmeter will reflect a corresponding change.

Effects of Changes in Amplitude on Controlled Devices

Extreme changes in amplitude will affect controlled devices. When the amplitude drops below the minimum operating voltage for a given device, that device will cease to function. When the amplitude exceeds the maximum safe operational voltage, the device can burn out. Other than these extremes, changes in amplitude will not affect a controlled devices operation. For example, changing amplitude will not alter the amount of fuel being sprayed through an injector.

Frequency

A far more important measurement to the automotive technician is frequency. Fre-

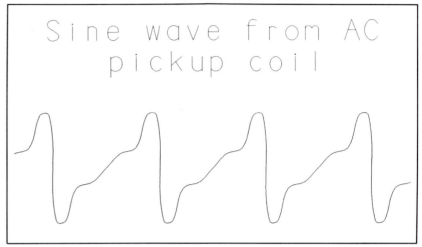

The pickup coil used in many ignition systems produces a sinusoidal waveform. If tested with an oscilloscope, the pickup coil should produce a signal similar to the pattern shown above.

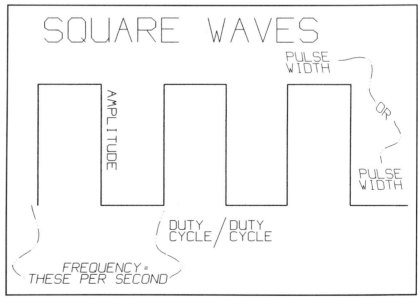

Hall Effects devices and optical sensors produce a square wave. The number of pulses created per second is called frequency. Change in voltage between the lowest part of the waveform and the highest part of the waveform is called amplitude and is measured in volts. The relationship between the part of the waveform where current is flowing and where current is not flowing is called duty cycle. Duty cycle is measured in percent. The length of time the current is flowing is called pulse width. The measurement of pulse width is in seconds or, usually, milliseconds.

quency is the number of pulses, or waveforms, that occur within a given period. The most common time interval is 1 second, the most common unit of measurement is Hertz, or cycles per second. Therefore, a square wave occurring ten times per second has a frequency of 10 Hertz. Note that a complete square wave includes both a high and a low voltage portion of the waveform.

There are three valid ways for an automotive technician to measure frequency: the frequency counter, oscilloscope, and tachometer.

Many electronic test equipment manufacturers produce dedicated frequency counters. These meters can be expensive and are usually capable of reading from fractions of a Hertz up to megahertz (millions of pulses per second). This level of flexibility is wasted on the automotive technician. Automotive electronics only requires an ability to read frequencies ranging from 1 to 500 Hertz.

To measure frequency on an oscilloscope, count the number of milliseconds required for the complete high and low voltage portions of the waveform. Divide the number of milliseconds into 1. For example, a square wave displayed on the oscilloscope covers 5 milliseconds.

1 divided by 0.005 seconds equals 200 Hertz

The tachometer may be the most accessible method for the automotive technician to measure frequency. However, there is a conversion formula required: rpm reading times the number of cylinders that the meter is set on (N) divided by 2 divided by 60.

$$[(RPM \times N) \div 2] \div 60$$

This formula can be abbreviated by always placing the tachometer on the four cylinder scale. On the four cylinder scale, simply divide the tach reading by 30. As an example, a tachometer is set on the four cylinder scale and connected to a circuit carrying a frequency. The tachometer displays 4300rpm.

$$4300 \div 30 = 143.33 \text{ Hertz}$$

How Changes in Frequency Affect Average Voltage in the Circuit

When the frequency changes, there is no change in the average voltage of a circuit. The exception to this is when the frequency drops to a point approaching the sampling frequency of the meter. At this point, the meter will begin to fluctuate with the pulsing and changes in frequency may become apparent.

Effects of Changes in Frequency on Controlled Devices

Changing the frequency of a signal to a controlled device will not alter the net result of that device. An injector will not spray more fuel or less fuel because of a change in frequency. A vacuum control solenoid will not alter the amount of vacuum applied to the EGR valve simply because of a change in frequency.

Duty Cycle

Duty cycle is the relationship between the high- and low-voltage portion of a square wave, measured in percentage. If the square wave has a high voltage half the time and a low voltage half the time, then the duty cycle is 50 percent.

Looking at a 50 percent duty cycle on an oscilloscope reveals a waveform with equal high- and low-voltage lines. If the waveform displayed on the scope had a high-voltage line that was one time unit long and a low-voltage line that was three time units long, then it is a little more difficult to ascertain the duty cycle. The duty cycle might be 25 percent or it might be 75 percent. The determining factor would be whether current is flowing through the circuit when the voltage is high or when the voltage is low. Therefore, looking at duty cycle another way, it is the relationship between current flowing through a circuit and current not flowing through a circuit measured in percent.

The Duty Cycle Meter

Many test equipment manufacturers offer a dedicated duty cycle meter. Others will offer a duty cycle function on voltmeters or engine analyzers. These meters offer a direct reading of duty cycle. The technician does not need to be concerned with when current is flowing through the circuit. The meter analyzes that.

Dwell Meter

Since the days of Charles Kettering (the man who developed the point/condenser distributor ignition system), we have been measuring the duty cycle of the primary ignition system (the more familiar term for this is dwell). Duty cycle is measured on a scale of 0 to 100. The four cylinder scale of the dwell meter reads 0 to 90. Using the dwell meter on the four cylinder scale will yield a displayed reading with only a 10 percent error. In most cases where the automotive technician is measuring duty cycle, an error of 10 percent is insignificant. If greater accuracy is required, the four cylinder scale reading can be multiplied by 1.1. For example:

36deg of dwell x 1.1 = 39.6 percent

Measuring duty cycle with a voltmeter can be accurate if the lowest voltage and the highest voltage of the square wave are both known. However, the calculation procedure is complex. In many cases there is no need to determine the duty cycle accurately, the need is to only see that the duty cycle is changing. As the duty cycle changes, the voltage on the voltmeter will change.

Effects of Changes in Duty Cycle on Average Voltage

As the duty cycle changes, the average voltage changes. As with determining duty cycle on an oscilloscope, the voltage may increase or decrease as the duty cycle increases.

How Changes in Duty Cycle Affect Average Voltage

The average voltage changes as the duty cycle changes.

How Changes in Duty Cycle Affect Controlled Devices

One common method automotive computers use to control the volume of something is through a variable duty cycle square wave. A variable duty cycle controlled solenoid operated valve is used to control fuel or vacuum. This valve could be either a normally open valve or a normally closed valve. If it is a normally open valve, then the flow through the valve decreases as the duty cycle increases. If the valve is normally closed, then the flow through the valve increases as the duty cycle increases.

Pulse Width

The measurement of pulse within automotive electronics is limited to measuring the length of time that the injector is open. Similar to duty cycle, it is the measurement of the length of time that current is flowing through the injector solenoid. The difference is that duty cycle is measured in percent, while pulse width is measured in fractions of a second, or actual time.

Pulse Width Meter

Electronics test equipment manufacturers build testers designed to measure the length of time that current is flowing through the injector. These meters tend to be expensive. Since it is rare that the measurement of injector width is important to troubleshooting, investing in a pulse width meter may not be wise.

Oscilloscope

To measure pulse width on an oscilloscope, it is only necessary to align the injector pulse with time measurement mark on the scope.

Tach/Dwell

For those not blessed with either an oscilloscope nor a pulse width meter there is an alternative. Measure the duty cycle and the frequency of the signal, then use the following formula: 1 divided by frequency times duty cycle percentage divided by 100.

For example: the frequency of the pulse is 25 Hertz; the duty cycle of the pulse is 3 percent.

$$(1 \div 25) \times (3 \div 100) = 0.0012$$
seconds or 1.2 milliseconds

How Changes in Pulse Width Affect the Average Voltage

The average voltage changes as the pulse width changes.

How Changes in Pulse Width Affect Controlled Devices

As the pulse width to the injector increases, the fuel flow through the injector also increases.

The Inductive Kick

When the current passing through a solenoid is shut off, the magnetic field created by that current surrounding the windings of the solenoid collapses. As the magnetic field collapses, it induces more voltage in the circuit. This induced voltage shows up as a spike on the square wave at the point where the current was shut off.

Types of Circuit Defects

There are three common defects in an electrical circuit. These defects are opens, shorts, and grounds.

Open

In an open circuit, the path to ground has been broken. When there is an open circuit there is no current flowing. Because there is no current flowing, there will be no voltage drops through the circuit up to the point of the open. Therefore, if the voltage is measured at the point of the open, source voltage will be indicated.

Short

This term is often misused. A short occurs when a conductor or device is bypassed through an alternate current path. This will cause the current flow in the circuit to increase, increasing the voltage available to the next device in series. The next devices in series, the power supply, and the current carrying conductor are at risk.

The term short is also correctly used when two conductors touch allowing an improper voltage from one conductor to be transferred into another conductor. Depending on the circumstances, current flow in the circuit may or may not be affected. Excessive voltage applied to a device may put that device at risk of being damaged.

Ground

This is the correct name for the condition that is often called a short. A ground occurs when the current path is directed back to the negative terminal of the battery prematurely, before the last powered device. This condition causes the current flow to increase, resulting in possible damage to the conductors and devices in the circuit. Voltage measured on the outbound side of the device in the series before the ground will be zero.

Parallel Robbery

When troubleshooting automotive electronic systems, each circuit being tested is treated as a series circuit. Many of the circuits tested are actually parallel circuits. A parallel circuit shares a common voltage source and a

17

common ground with another circuit. By this definition, all circuits on an automobile are technically parallel circuits. Current will always take the easiest path to ground. If the resistance in one leg of the parallel circuit drops, it will tend to pull power (current) away from the other leg. Thus, problems in one leg of the parallel can affect the operation of all the other legs.

Electrostatic Discharge (ESD)

Since the early eighties, much has been said about the effects of static electricity on electronic components. Some components are more sensitive than others. Semiconductor devices are what all the furor has been about. There are several different types of semiconductors used in automotive electronics, each with a different degree of sensitivity.

Examples of ESD Vulnerability

Circuitry	Applications	Voltage
MOS circuitry	few automotive applications	1 to 1000 volts
CMOS circuitry	Ford EEC I, II, III	1000 to 5000 volts
TTL circuitry	most automotive systems	5000 to 15000 volts

At first, this looks like a large amount of voltage to be sent into an electronic device accidentally. Following are some examples of static voltages generated by everyday activities.
- Stroking a rubber comb through dry hair -2500 volts
- Rolling a desk chair across a plastic floor mat 2000 volts
- Crumpling a polyethylene bag 300 volts
- Rubbing a pencil eraser across a circuit board +100 volts
- Rapidly unrolling black electrical tape +500 volts
- Walking across a carpet with rubber sole shoes -1000 volts

The previous examples are at 75deg F and 60 percent relative humidity. Drop the relative humidity to 10 percent and these numbers can be up to fifty times higher.

To Protect from ESD:
- Store components in their original packages.
- Do not touch component leads or pins.
- Before handling the computer, discharge the static from your body by touching a grounded metal surface.
- Do not slide the computer or device across any surface.
- Keep static-generating materials (such as plastic, cellophane, paper, candy wrappers, and cardboard) away from the work area.
- Keep clothing away from static sensitive devices.
- Never connect static sensitive devices with power applied.

Subunits of Automotive Electronics Measurement
Volts
1 microvolt = 0.000001 volt
1 millivolt = 0.001 volt
1 kilovolt = 1000 volts

Amps
1 microamp = 0.000001 amp
1 milliamp = 0.001 amp

Ohms
1 kilo-ohm = 1,000 ohms
1 megohm = 1,000,000 ohms

Watts
1 milliwatt = 0.001 watt
1 kilowatt = 1000 watts

Pickup Coil

A more scientific term for this device is variable reluctance transducer (VRT). The typical automotive technician will think of the electronic ignition distributor when this component is mentioned. In reality, there are many uses for this device. The VRT is used to measure crankshaft rotational speed on some distributorless ignition systems, wheel speed for four-wheel antilock braking systems, differential speed on rear wheel antilock braking systems, and, in some vehicles, speed sensors. The VRT produces an AC sine wave with a frequency directly proportional to the speed of the rotation.

The primary advantage of the VRT over other rotational speed sensors is its simplicity. Consisting of a coil of wire, a permanent magnet, and a rotating reluctor, there is very little that can go wrong with it. Its main disadvantage is its inability to accurately detect low speed rotation (at low rotational speeds, the VRT is unable to produce a signal).

How the VRT Works

A coil of wire sits in a magnetic field created by a permanent magnet. A metal wheel with protruding reluctor teeth rotates through the magnetic field. As it rotates and one of the teeth approaches the magnetic field, the magnetic field is bent toward the approaching tooth. As it is bent, it passes across the coil of wire inducing a voltage. Continuing to rotate the reluctor tooth drags the magnet field across the coil of wire eventually bending it in the opposite direction. The result is an AC signal.

Testing the VRT
With an Ohmmeter

Most books on troubleshooting electronic ignition systems will suggest using an ohmmeter to test a reluctance pickup. Although this is not a totally invalid test, it only tests the coil of wire. Reluctor air gap, condition of the permanent magnet, and adequate rotational speed are not tested with this method.

Disconnect the pickup from the ignition module or vehicle wiring harness leads. Connect the ohmmeter to the pickup coil leads and measure the resistance. Typical ohmmeter readings for a good reluctance pickup coil would be between 500 and 1,500 ohms.

With an Oscilloscope

A much better way to test a reluctance pickup is with an oscilloscope. Connect the scope to the pickup coil leads and rotate the reluctor (crank the engine, rotate the wheel). A series of ripples should appear on the scope. If the line remains flat there is an open in the coil of wire, the permanent magnet has been damaged, or the air gap between the pickup and the reluctor teeth is too large.

With an AC Voltmeter

The AC voltmeter is a practical and effective alternative to the oscilloscope. Connect the AC voltmeter in the same manner as described for connecting the oscilloscope. Rotating the reluctor at minimum speed (cranking the engine, rotating a wheel at one revolution per second) would yield between 0.5 and 1.5 volts. If the AC voltmeter does not produce a voltage, there is an open in the coil of wire, the permanent magnet has been damaged, or the air gap between the pickup and the reluctor teeth is too large.

Notes on Testing. Since the VRT is primarily a coil of wire and a permanent magnet, it is prone to intermittent failures with changes in temperature and vibration. If the failure being diagnosed is intermittent, the sensor should be heated and tapped while testing.

Hall Effects Sensor

The Hall Effects Sensor is often used as an alternative to the VRT. Many ignition systems, both distributorless and distributor type, use a Hall Effects device. Its primary advantage over the VRT is its ability to detect position and rotational speed from zero rpm to tens of thousands. Its primary disadvantage is that it is not as rugged as the VRT and is more sensitive to errant magnetic fields. An intense magnetic field can shut down the proper operation of a Hall Effects.

How the Hall Effects Works

A Hall Effects pickup is a semiconductor carrying a current flow. When a magnetic field falls perpendicular to the direction of that current flow, part of that current is redirected perpendicular to the main current path. The semiconductor is placed near a permanent magnet. A set of metal blades, or armature, attached to a rotating shaft or other device passes between the hall effects semiconductor and the permanent magnet. As the armature rotates, the magnet field is alternately applied to the Hall Effects and interrupted. The result is a pulsing current perpendicular to the main current path. This frequency is directly proportional to the speed of armature rotation. Since the output is only dependent on the presence of the magnetic field, the Hall Effects unit is capable of detecting armature position even when there is no rotational speed.

Testing the Halls Effects
With an Ohmmeter

There is no valid test procedure on the Hall Effects using an ohmmeter.

With an Oscilloscope

Connect the oscilloscope to the Hall Effects signal lead. Rotate the armature. Depending on the number of blades and the rotational speed of the armature, the scope pattern could appear either as a square wave or a flat line that rises and falls with rotation.

With a Voltmeter

Connect a voltmeter to the Hall Effects output lead. The voltmeter should display either a digital high (4 volts or more) or a digital low (around 0 volt). Slowly rotate the armature while observing the voltmeter. If the volt-

The AC pickup coil is the oldest and most dependable of the three devices commonly used to measure shaft rotation in an electronic ignition system. It is easily recognized because it has only two wires; the optical sensor and Hall Effects will have at least three.

Next to the pickup coil, the most common shaft speed/position sensor is the Hall Effects. Ford and Chrysler use the Hall Effects exclusively on their fuel-injected engines. More accurate than the pickup coil, its output signal requires less modification by a fuel injection computer.

meter had read low it should now read high; if the voltmeter had read high it should now read low. If the voltage fluctuates in this manner as the armature is rotated, then the Hall Effects is good.

With a Dwell Meter

Since the signal generated by the Hall Effects is a square wave, the dwell meter becomes a natural for testing. Connect the dwell meter between the Hall Effects output and ground. Rotate the armature as fast as possible (example, crank the engine). The dwell meter should read something besides zero and full scale. If it does, the Hall Effects is good.

With a Tachometer

As with the dwell meter, the tachometer is also a good tool for detecting square wave. Connect the tachometer between the Hall Effects output and ground. With the armature rotating as described in the paragraph on the dwell meter, the tachometer should read something other than zero if the Hall Effects is good.

Optical Sensor

The signal produced by the optical sensor is identical to the one produced by the Hall Effects. The signal, however, is produced by an armature interrupting light. A light emitting diode (usually infrared, invisible light) sits opposite an optical receiving device such as a photodiode or phototransistor. An armature is rotated between the LED and the receiver; unlike the Hall Effects the armature in the optical sensor can be metal, plastic, or any translucent material. As the armature rotates, light alternately falls on and is kept from falling on the receiver. As this occurs, the current flowing through the receiver is turned on and off creating a square wave with a frequency directly proportional to armature rotation. In some cases, such as many GM vehicle speed sensors, the light is reflected off of rotating blades rather than interrupted.

The main advantage of the optical rotational sensor over the VRT and the Hall Effects is its ability to produce extremely high frequencies. The 3.0 liter Chrysler distributor produces a frequency of 540,000 Hertz (0.54 megahertz!) at just 3000rpm. The primary disadvantage is a sensitivity to dirt, oil, and grease creating erroneous signals.

Testing the Optical Sensor
With an Ohmmeter

There is no valid test procedure on the optical sensor using an ohmmeter.

With an Oscilloscope

Connect the oscilloscope to the optical sensor signal lead. Rotate the armature. Depending on the number of blades and the rotational speed of the armature, the scope pattern could appear either as a square wave or a flat line that rises and falls with rotation.

With a Voltmeter

Connect a voltmeter to the optical sensor output lead. The voltmeter should display either a digital high (4 volts or more) or a digital low (around 0 volt). Slowly rotate the armature while observing the voltmeter. If the voltmeter had read low, it should now read high; if the voltmeter had read high, it should now read low. If the voltage fluctuates in this manner as the armature is rotated, then the optical sensor is good.

With a Dwell Meter

Since the signal generated by the optical sensor is a square wave, the dwell meter becomes a natural for testing. Connect the dwell meter between the optical sensor output and ground. Rotate the armature as fast as possible (example, crank the engine). The dwell meter should read something besides zero and full scale. If it does, the optical sensor is good.

With a Tachometer

As with the dwell meter, the tachometer is also a good tool for detecting square wave. Connect the tachometer between the Hall Effects output and ground. With the armature rotating as described in the paragraph on the dwell meter, the tachometer should read something other than zero if the optical sensor is good.

The Principles of Spark Ignition Combustion

The spark ignition engine uses the heat of an electrical spark to ignite the air/fuel charge in the combustion chamber. The heat from two other sources assists the spark in the ignition process. Heat that is generated in the spark ignition engine as the piston travels upward on the compression stroke is called adiabatic heating. Some of the heat of combustion is retained in the metal walls of the cylinder. When the heat from either of the other two sources is reduced, the ignition system must work harder.

The engineers say that it requires about 0.2 mJ (microjoules) of energy to ignite a stoichiometric mixture. Stoichiometric means the air and fuel have been combined in the combustion chamber in a ratio of 14.7 weight units of air to 1.0 weight units of fuel. When the engine is running lean, the amount of energy required to ignite the mixture can climb as high as 3.0 mJ.

To put this information into perspective, one Joule is equal to 0.0009478 Btu's (British thermal units). Therefore, the amount of heat energy required to ignite the air/fuel mixture is 0.00018956 Btu's. Compare this to small room heaters that produce several thousand Btu's. The amount of energy required to ignite the mixture is small. Such a small amount of energy seems easy to produce. The trick is to produce this energy within 2.5 milliseconds and within a narrow time window. If the engine is running at 3600rpm, the crankshaft is rotating 60 times per second. At this speed, it takes about 0.00005 seconds (0.05 milliseconds) for the crankshaft to rotate 1deg. This gives a window of approximately 0.4 milliseconds for the spark to be initiated. This is the real trick.

The heart of the spark ignition system is the ignition coil. An ignition coil is a step-up transformer. Like other transformers the ignition coil contains two sets of windings. A large, short set of windings, called the primary, carries a low-voltage current from the vehicle's electrical system. Although in many ignition systems this voltage is dropped through a ballast resistor, we will refer to this part of the ignition as the 12 volt or primary circuit. This set of windings generates a magnetic field.

A thin, long set of windings will have a high-voltage current induced when the magnetic field generated by the primary collapses. To collapse the magnetic field, the flow of current through the primary windings must be interrupted. From the 1920s into the 1970s, the current flow was interrupted by opening a cam operated switch called points. Since the cam operating the points is geared to the engine camshaft, the opening and closing of the points is perfectly timed and synchronized.

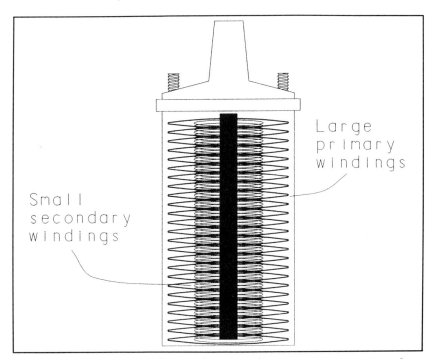

The ignition coil consists of a set of large, high current flow winding called the primary windings. Through the primary passes 2-5 amps of current. When the current is shut off by the opening of the ignition points, or by the ignition module, a magnetic field collapses, inducing a voltage in the much smaller, but much more numerous secondary windings. Because of the ratio in the number of primary turns versus the number of secondary turns, a high voltage is induced in the secondary. For decades, both sets of winding were submerged in oil to keep them cool.

There are two main problems with points. First, they are a dirty switch. This means that when the points open a potential for arcing is extremely high. When the points arc, the collapse of the magnetic field in the primary becomes erratic, dissipating and wasting some of the potential energy. The second problem is that they are subject to mechanical wear.

The primary windings of the ignition coil in the classic point/condenser ignition system carry a current flow of about 2.5 amps when the points are closed. A ballast resistor located between the coil and the ignition switch creates a voltage drop of about 9.6 volts across the coil. This means that there is a potential of 24 watts of power in the primary. When the points open, the magnetic field of the primary collapses across the secondary. This induces a potential of 24 watts of power in the secondary. The coil, however, is an imperfect transformer and some of this potential power will be lost. If all this power were to be used in an instant across the gap of the spark plug and if the voltage required to jump that gap were 8,000 volts (a typical real-world voltage), then the amperage carried in that spark would be 0.003 amps. However, the spark must be maintained across the gap for between 1.5 and 2.5 milliseconds.

Initially, the voltage will be quite high, the 8,000 volts or more. This voltage is necessary to initiate the spark across the gap of the spark plug. This is an ionizing process. Once the gap is ionized, the voltage required to maintain the spark across the gap drops rapidly. The voltage required to maintain the spark across the gap is only about 1,500 volts. After 1.5 to 2.5 milliseconds, the energy required to maintain the spark across the gap has been exhausted and the spark collapses.

The advantage of many modern high energy style electronic ignition systems is found in the primary side of the ignition system. Most of these systems boast a current flow of 5 amps or more. Also, the voltage drop across the ignition coil primary is a full 12.6 volts or more. This means there is 63 watts of power in the primary. This is 2.6 times greater potential energy than is available in the standard point/condenser ignition. As a result, at 8,000 volts there is 0.008 amps available to jump the gap of the spark plug. This is important when there are problems in the condition of the engine, ignition, or air/fuel ratio.

As the condition of the engine deteriorates, problems with the valves and rings for instance, the compression drops. Normally, as the intake valve closes and the piston comes up on the compression stroke, the temperature of the air/fuel charge in the combustion chamber rises dramatically. Just prior to ignition, the adiabatic heating that is a result of compression will raise the temperature of the air by several hundred degrees Celsius. This heating becomes an essential accompaniment to the spark. When the compression drops due to engine wear, the ignition must work harder. The high energy electronic ignition system provides 2.6 times greater potential energy to meet this need.

As previously stated, a stoichiometric air/fuel ratio will require one-fifteenth the ignition power of a lean mixture. The high energy electronic ignition

Most modern ignition coils are not oil filled. Heat is displaced through the plastic material surrounding the primary and secondary windings.

Thousands of volts are required to jump the gap at the tip of the spark plug. As the gap grows larger, either through use or due to improper gapping by the technician, the voltage across the gap increases. For the bulk of the time the spark is jumping the gap, there is only 1,000 to 2,000 volts. However, many times that amount is required to begin the spark jumping the gap.

system will be able to successfully fire a much leaner mixture than a point/condenser ignition system.

Secondary ignition parts wear as they are used. The extra available power created by the high energy electronic system enables the ignition system to fire even though the secondary ignition components are badly worn.

The ability of the ignition system to fire the cylinder in spite of these three factors points out the reduced necessity for maintenance and service. Increased maintenance intervals, reduced maintenance costs and better fuel economy are all advantages of electronic ignition systems.

Fuel injection, tough emission standards, and extended service intervals have not eliminated the need for regular maintenance and routine tune-ups. Many cars of the late eighties and even the nineties boast of spark plug replacement at intervals of 50,000 miles or more. There are still air filters and fuel filters with similar service intervals. These service intervals may be fine for typical consumers who use their cars to go to and from work with an annual trip to Yellowstone or Disneyland. Owners who expect a little more performance, economy, or reliability out of their cars should consider the following tune-up maintenance schedule.

• Every 12,000 miles: replace spark plugs, fuel filter, and air filter; check vacuum hoses and intake air tubes.
• Every 24,000 miles: replace spark plugs, fuel filter, distributor cap, rotor, plug wires; check vacuum hoses and intake air tubes.

Using the correct spark plug heat range is critical. It is inadvisable to use any heat range other than the original unless a manufacturer's service bulletin announces an advised change. The spark plug on the left is a colder plug than that on the right.

The Story of the Spark Plug

Replacing a car's spark plugs is one of those tasks that gives every Saturday afternoon mechanic a feeling of pride. Yet, there are many subtle details about spark plug replacement and much that an old spark plug can tell you. Let us begin with some of the basics of what the spark plug does.

The spark plug ignites the air/fuel charge in the combustion chamber. It accomplishes this by a high-voltage spark across its electrodes. The source of the spark is the ignition coil. The spark is conducted to the spark plugs by the secondary ignition cables. This spark lasts for about 1.5 to 2.0 milliseconds at around 1,000 volts. However, for the first 30 microseconds, the voltage of the spark is much higher, somewhere between 5,000 and 30,000 volts. This higher voltage is necessary to initiate, or begin, the spark across the gap of the plug. The ignition coil is basically a transformer. The amount of energy available to the plugs is limited to the number of watts passing through the primary. If too much of the available energy is used to initiate the spark, then the amount of energy left to maintain the spark will be re-

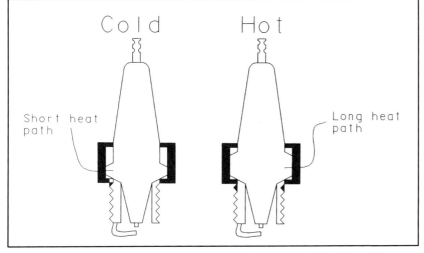

Although tip protrusion can be a clue to the relative heat range of a spark plug, the real determining factor is how long the heat path is. This cannot be determined externally.

duced.

Today's leaner running engines will extinguish the fire in the cylinder when the spark goes out. Therefore, anything that affects the duration of the spark affects power and driveability much more than it did in the sixties. Anything that affects the voltage required to initiate the spark will affect the duration of the spark.

The following will affect the spark initiation voltage:
• Spark plug heat range
• Compression (and compression ratio)
• Air/fuel ratio
• Spark plug wear
• Condition of distributor cap
• Condition of the ignition rotor
• Condition of the ignition coil
• Condition of the secondary ignition wires

Heat Range

Unless you have a highly modified engine, you should stick with the heat range recommended by the car spark plug manufacturer. If the incorrect heat range selected is too cold, it can increase the spark initiation voltage, robbing power from the engine. Another difficulty with a cold plug is that it can result in misfiring and fouling. Using spark plugs with too hot of a heat range can result in pre-ignition or pinging.

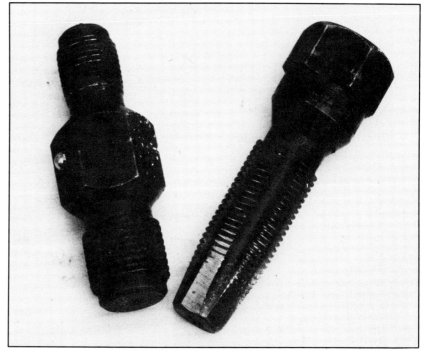

Even when great care is taken when removing and installing the spark plugs, the threads can be damaged. For just a few dollars, thread chasers can repair the damage. When the damage is more severe, a thread replacement such as a helicoil may be required.

Compression

The effect of compression on the power from an engine is like a two-edged sword. On the one hand, increased compression increases the power potential of the engine; yet increased compression also increases spark initiation voltage. The increase in spark initiation voltage means that high performance, high compression engines may require an ignition coil with greater potential energy. As compression in each cylinder drops, the voltage required to initiate the spark across decreases; but so does engine power.

Air/Fuel Ratio

Air/fuel ratio is one of the most critical variables when it comes to driveability. If the air/fuel ratio is too rich, spark initiation voltage will be quite low and the burn in the combustion chamber will be slow and incomplete. A lean air/fuel ratio makes it difficult for the spark to start jumping the spark plug gap. A great deal of energy will be used during the first 30 microseconds leaving little energy to maintain a burn in the combustion chamber. The result is

A handy but obsolete tool is this flexible dwell adjusting tool. Old-style GM distributor caps had a window in the side that allowed the technician to adjust the dwell while the engine was running.

reduced power, stumbling, hesitation, and misfire.

Plug Wear

If you have ever removed an old spark plug, you may have noticed that the gap is much wider than when it was installed. This is because, like many other components in the engine, spark plugs wear as they are used. The wider the gap, the higher the voltage required to begin the spark jumping the gap, and the less energy there is to accomplish the burn in the combustion chamber. This has a direct effect on combustion; decreasing power, increasing toxic emissions, and causing poor driveability.

Condition of the Distributor Cap, Rotor, and Plug Wires

Within the high-voltage side of the ignition system there are several other items that wear as they work. The distributor cap, on cars equipped with a distributor, is used along with the rotor to transfer the high-voltage spark from the ignition coil to the spark plug wires. High voltage arrives at the center of the distributor cap either through the coil wire or directly from the ignition coil, this voltage is conducted to the rotor by a carbon nib. The current then travels through either a solid metal conductor or a resistive element in the rotor. As the rotor swings past each of the plug wire electrodes, a spark jumps from the rotor. Wear on the carbon nib or an excessive gap between the rotor and the distributor cap electrodes will increase the amount of energy consumed on its way to the spark plugs. This decreases the amount of available energy to fire the cylinder. Distributorless ignition systems are one of the methods used to reduce the number of places where wear can affect secondary ignition energy thereby increasing tune-up service intervals. As plug wires age, their resistance increases and as their resistance increases a greater portion of the potential spark energy heading toward the spark plugs is consumed.

**Replacing the Spark Plugs
Testing**

There are two ways of testing spark plug wires. The first is with the use of an engine analyzer oscilloscope. The second method is with the use of an ohmmeter. Remove each plug wire, set the ohmmeter on the x1000 ohm scale, and measure its resistance end to end. A good plug wire will have less than 10,000 ohms but greater than 1,000 ohms per foot.

Removal

With the engine cold, unscrew the spark plugs two or three turns. If a spark plug is difficult to unscrew, it might have dirt or grit on the threads. Put a couple of drops of oil on the exposed threads and allow it to soak in for a few minutes. Screw the plug back in, then out several times, add a little more each time until the plug is removed. Using compressed air or a solvent soaked brush, clean the area around each plug to remove any dirt or foreign objects that might have fallen into the cylinder when you removed the plugs. Inspection of the removed spark plugs can tell you much about the running condition of the engine. Carbon deposits, ash formations, oil fouling, soot, and other conditions can be indicative of engine or injection system problems.

Installation

Before installing the new spark plugs, be sure to check the gap. Although some spark plug manufacturers make an effort to pre-gap their plugs, they can be unintentionally re-gapped in shipping. Inspect the threads in the cylinder head to insure that they are clean and undamaged. Also, check the mating surface where the plug contacts the head, it should be clean and free

The plug wire retainers are one of the first components in an ignition system to deteriorate. Many technicians simply throw the broken retainers in the trash and allow the wires to fall where they may. In the days of massive chrome bumpers, this was all right. Today, induced voltages from misplaced plug wires can fool the fuel injection system and cause the engine to run either extremely rich or extremely lean.

of burrs. Start the new spark plug and screw it in a few turns by hand. Continue to screw the plug in either by hand or with a socket and ratchet until it contacts the mating surface firmly. To avoid overtightening, use a torque wrench and tighten a tapered seat plug to 26ft/lb in a cast iron head or 21ft/lb in an aluminum head. In the real world, a torque wrench is seldom used when installing spark plugs so use this rule of thumb: firm contact plus 15deg of rotation.

Replacing Secondary Ignition Cables, Cap, and Rotor

There are only a couple of important things to remember when replacing secondary ignition wires. First, if the plug wires are installed in the incorrect order, a backfire may occur resulting in damage to the air flow meter, air mass meter, or the rubber tube that connects it to the throttle assembly. The original equipment plug wires have numbers on them indicating to which cylinder they should be connected. Aftermarket or replacement plug wires may not have these numbers.

3M and other companies make adhesive numbers that you can attach to insure proper reinstallation.

Replacing the Distributor Cap

Distributor cap replacement consists of releasing the attachment screws or clips and making sure that the plug wires are installed in the correct order.

No matter what type of distributor you have on your car, it is a good idea to replace the distributor cap and rotor together and use the same brand. Pairing caps and rotors of two different

No matter how good the ignition system is, it cannot compensate for a damaged, degraded, or worn-out engine. When the author ran and owned a repair shop, he required that every tune-up include a compression test. This was done at no charge to the customer, but it allowed the technicians to predict the effectiveness of the tune-up.

Air/fuel ratio is critical. As a cylinder runs leaner, the amount of energy required to jump the gap across the tip of the spark plug increases. Soon the limit of the ignition system is reached and a misfire occurs. Restricted injectors or misadjusted carburetors can result in a misfire that seems to be ignition related.

Although for many applications the rotor electrode is a solid conductor that runs from the center to the end, this electrode has a resistor in the middle. Most rotors are cheap and should be replaced during every tune-up.

Although seemingly insignificant, there are two problems in this distributor cap that could contribute to the quality of ignition. First, the burned areas on the edge of each electrode represents a point of resistance. As the resistance between the rotor and the cap increases, the amount of voltage and energy required to make the connection between the cap and rotor also increases, potentially resulting in a misfire. The second problem is the oily film and dirt on the inside of cap. This could supply an alternative current path for the spark and result in a misfire.

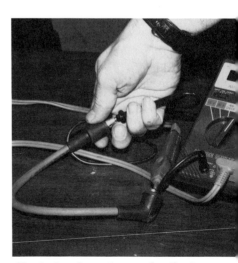

Check the plug wires and coil wire with an ohmmeter. The resistance must be less than 15,000 ohms per foot. This coil wire is about 10in long and has a resistance of 4.37 ohms.

manufacturers can result in the incorrect rotor air gap. Excessive rotor air gap can cause excessively high spark initiation voltage and can result in incomplete combustion.

Replacing the Air Filter

The real value of the air filter is significantly underestimated. It is the engine's only defense against sand, grit, and other hard particle contamination. When these substances enter the combustion chamber, they can act like grinding compound on the cylinder walls, piston rings, and valves. Replace the air filter at least once a year or every 24,000 miles. In areas where sand blows around, like western Texas or Arizona, the air filter should be replaced much more often.

On most carbureted cars, a restricted air filter will cause the engine to run rich. This is because the restriction causes a reduction of pressure in the venturi of the carburetor while the pressure in the fuel bowl remains constant at atmospheric. This increased pressure differential increases the flow of fuel into the venturi and the mixture enriches. On a fuel injected engine, the electronic control unit (ECU) measures the exact volume of air entering the engine by the air flow meter and delivers the correct amount of fuel for the measured amount of air. Restrict the flow of incoming air, less air will be measured, less fuel will be metered into the engine. D-Jetronic air measurement is not precise and therefore may run a little rich as a result of a restricted air filter.

Replacing the Fuel Filter

The fuel filter is the most important service item among the fuel components of the fuel injection system. In my seventeen years of experience with fuel injected cars, I have replaced many original fuel filters on cars that were over ten years old. This lack of routine maintenance is just begging for trouble.

After removing the fuel filter, find a white ceramic container, such as an old coffee cup, and drain the contents of the filter through the inlet fitting. Inspect the gasoline in the cup for evidence of sand, rust, or other hard particle contamination. Now pour the gas into a clear container such as an old glass and allow it to sit for about thirty minutes. If there is a high water content in the fuel, it will separate while sitting. The fuel will float to the top. If there is excessive water or hard particle contamination in the tank, it may have to be removed and professionally cleaned. For minor water contamination problems, you can purchase additives at your local parts store.

Should the fuel filter become excessively clogged, the following symptom might develop. You start the car in the morning and it runs fine. As you drive several miles, the car may buck a little or loose a little power, then suddenly the engine quits almost as though someone had shut off the key. After sitting on the side of the road for several minutes, the car can be restarted and driven for a couple of miles before the symptom recurs. A severely restricted fuel filter could cause this problem.

The car runs well initially because the bulk of what is causing the restriction has fallen to the bottom of the fuel filter as

Replacement plug wires range in price from a few dollars a set to more than $50. Like many things, you get what you pay for. Cheap wires usually last only a short time and may cause radio static. On electronic fuel-injected cars, cheap plug wires can interfere with the proper operation of sensors.

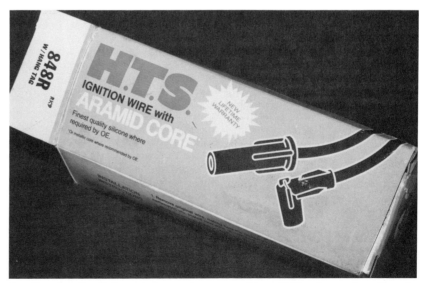

Overheard at South Loop Auto supply in the mid-seventies: "...if you had a team of eight horses and one broke a leg, you wouldn't shoot all eight would you?" For those who adhere to the financial benefits of this philosophy, plug wires are sold in single wire packages.

One of the author's former customers owned a Subaru that had survived Washington state's 1980 Mount St. Helens eruption. For over a year, the customer had taken the car to technicians for a lack of power. The spark plugs, plug wires, points, condenser, distributor cap, and rotor had been replaced several times. No one had ever replaced the air filter. No matter how good ignition parts are, if the engine cannot get air it will not run right.

Every tune-up should include the replacement of the fuel filter. In the days of the carburetor, fuel filters were designed to pass relatively large particles without problems. Today's fuel injected engines have fine filters that pass only the tiniest particles. Replace the fuel filter during each tune-up.

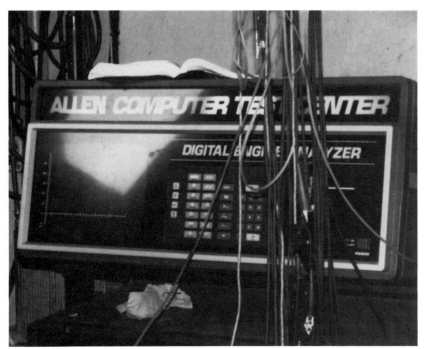

The engine analyzer is a useful piece of equipment. Although essential to the professional technician, the price of such analysis is seldom justified by the information gained. In most cases the cost of the analysis is on a par with the price of the parts it is best at diagnosing. The typical operator of this equipment can recognize defective plugs, ignition wires, distributor caps, and rotors. For many applications the do-it-yourselfer can replace these parts for less than the price of the test.

sediment. When the engine is started and the fuel begins to flow through the injection system, this sediment gets stirred up and pressed against the paper elements of the fuel filter. As it does, fuel volume to the injectors is decreased and the engine begins to run lean. Sooner or later the engine leans out so much that it dies. When the fuel filter becomes that restricted, some of the contaminants have forced their way through the filter and may contaminate the rest of the fuel system.

Alcohol Contamination

Alcohol contamination can damage many of the fuel injection components. Use of gasohol can be one source of alcohol. If you suspect that excessive alcohol content has caused the failure of system components, test a fuel sample for alcohol content. Pour 200 milliliters of the sample fuel into a glass or clear plastic container along with 100 milliliters of water. Immediately after putting the two liquids in the container, the dividing line will be at the 100 milliliter mark. Wait about thirty minutes, if the dividing line rises by more than 10 percent of the volume of the contents of the container then there is excessive alcohol in the fuel. Drain the fuel from the tank and replace it with good fuel.

There are many fuel additives on the market which contain alcohol. The alcohol content of these additives is so small compared to the size of the typical fuel tank that it poses no threat to the fuel system. Nevertheless, use caution, ask around a bit, and be selective when purchasing these products, some are much better than others.

A Rich Running Condition Caused by a Tricked Lambda Sensor

A Lambda sensor is tricked by:
- an exhaust leak
- low compression
- a secondary ignition problem
- incorrect ignition timing
- a defective air pump system
- a defective Lambda sensor

Testing for the Dead Hole

When a late-model fuel injected engine has a cylinder with a misfire, the effects can go far beyond a rough idle or loss of power. A cylinder that is still pulling in air but not burning that air will be pumping unburned oxygen past the Lambda sensor. This confuses the ECU, making it believe that the engine is running lean. The ECU responds by enriching the mixture and the gas mileage deteriorates dramatically.

You can use several effective methods to isolate a dead cylinder. All of these methods measure the power produced in each cylinder by killing them one at a time with the engine running a little above curb idle.

Back in the good old days, we used to take a test light, ground the alligator clip, and pierce through the insulation boot at the distributor cap end of the plug wire. This ground out the spark for one cylinder and an rpm drop would be noted. The greater the rpm drop, the more power that cylinder was contributing to the operation of the engine. Actually this is a valid testing procedure; however, piercing the insulation boot is only asking for more problems than you started with.

Another previously used method of performing a cylinder balance was to isolate the dead hole by pulling off one plug wire at a time and noting the rpm drops. The problem with this method is that you run the risk of damaging either yourself or the ignition module with a high-voltage spark. Let's explore some alternatives.

Cylinder Inhibit Tester

Several tool companies produce a cylinder shorting tach/dwell meter. These devices electronically disable one cylinder at a time while displaying rpm. Engine speed drops can be noted. Unfortunately, these testers can cost $500 or more.

Another method does the old test light technique one better. Cut a piece of 1/8in vacuum hose into four, six, or eight sections each about 1in long. With the engine shut off and one at a time, so as not to confuse the firing order, remove a plug wire from the distributor cap, insert a segment into the plug wire tower of the cap, and set the plug wire back on top of the hose. When you have installed all the segments, start the engine. Touching the vacuum hose conductors with a grounded test light will kill the cylinder so that you can note rpm drop. Again, the cylinder with the smallest drop in rpm is the weakest cylinder.

Whichever method you use, follow this procedure for the best results.

1. Adjust the engine speed to 1200 to 1400rpm by blocking the throttle open. Do not hold the throttle by hand, you will not be steady enough.

2. Electrically disconnect the idle stabilizer motor to prevent its affecting the idle speed.

3. Disconnect the Lambda sensor to prevent it from altering the air/fuel ratio to compensate for the dead cylinder.

4. Perform the cylinder kill test; rpm drop should be fairly equal between cylinders. Any cylinder that has a considerably less rpm drop than the rest is weak. Proceed to step 5.

5. Introduce a little propane into the intake, just enough to provide the highest rpm. Repeat the cylinder kill test. If the rpm drop from the weak cylinder tends to equalize with the rest, then you have a vacuum leak to track down. If the car has an idle stabilizer, be sure to disable it so that it will not attempt to compensate for the lack of power input from the dead cylinder.

The Point/Condenser Ignition System 3

The point/condenser ignition system served as the workhorse ignition system of the gasoline engine for over fifty years. Charles Franklin Kettering designed the system in the early part of the twentieth century.

Primary Ignition Components
Coil

The heart of any spark ignition system is the coil. Basically, the ignition coil used in the point/condenser ignition system is a step-up transformer inside of an oil filled metal can. As there are many times more secondary windings as primary windings, the voltage output of the secondary is potentially tens of thousands of volts.

Although the coil is relatively trouble-free, both the primary and secondary windings are subject to opens, shorts, grounds, and corrosion. When I was in trade school, and I confess it was in the days when electronic ignition was viewed as a gimmick and aberration, my tune-up instructor pointed out the valid telltale sign of a coil defect. Most point/condenser ignitions have a recess in the bottom similar to an aluminum soda can. An internal problem in the coil can cause overheating, which can cause the oil intended to cool the coil to expand, pushing the recessed area at the bottom of the coil out. Although this evidence is not conclusive as to coil condition, one way or the other, it can be one piece of the diagnostic puzzle.

Ignition Switch

The ignition switch is normally a key lock switch that serves as an on-off switch for primary current flow. In the typical point/condenser ignition system, there are two circuit paths through the switch for the primary side of the coil. The first path carries current to the ignition coil through a ballast resistor that reduces the voltage drop across the coil primary to about 9.6 volts and limits the current

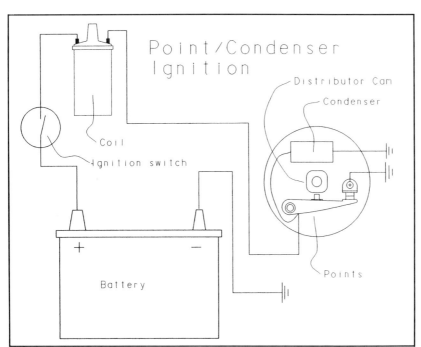

The point/condenser ignition system was the standard of the industry for many decades. However, the points required regular replacement. As the rubbing block on the points wore, the dwell and timing would change. This meant the ignition system was unable to maintain proper adjustments for low emissions. Tough emission standards meant elimination of the point/condenser system.

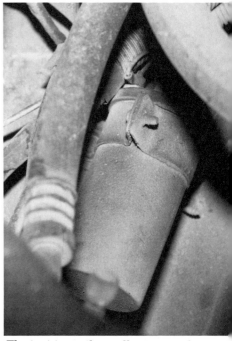

The ignition coil usually sits quietly in an obscure part of the engine compartment. Here it quietly does its job. The oil fill coils, like this coil on a 2.2 liter Dodge would often outlast the vehicle.

flow to about 2.5 amps. The second path carries current directly to the ignition coil when the engine is being cranked for starting. Bypassing the ballast resistor increases the current flow and the available secondary power while starting the engine. This is especially important when the engine is being cold started.

The ignition switch is one component of the ignition system that is often overlooked during the troubleshooting procedure. Although seldom the cause of driveability problems, an intermittent open circuit in the ignition switch can cause stalling and power and performance related problems.

Battery

Most people, including most professional automotive technicians, do not think of the battery as part of the ignition system. Yet, the battery is the ultimate power source for every electrical and electronic component on the car. An intermittent open circuit in the battery can manifest itself in a wide variety of symptoms, including intermittent coil operation, which can cause stalling and misfiring.

An important, but often ignored, part of the primary ignition system is the ignition switch. High resistance in an old ignition switch can limit the power potential of the coil. When most people think of the ignition switch, they think of the part they put the key into. However, in most modern application this is only a lock assembly.

Ballast Resistor

The ballast resistor is a low resistance, high-wattage resistor that is installed in the primary circuit to limit current flow. In addition to limiting current flow, it also causes reduction in the voltage drop across the ignition coil primary. The limiting of current and voltage reduces wear on the points by reducing their tendency to arc as they are opened to interrupt primary cur-

Mounted on the steering column is the electrical part of the ignition switch. The switch so seldom causes problems that it is often not considered during a driveability diagnosis. The ignition switch is as important a part in the primary ignition system as is the ignition coil.

The job of the condenser is to absorb the energy of self induction created when the magnetic field generated by the primary collapses across the primary. A faulty condenser can result in arcing at the ignition point and erratic engine operation.

The ignition points are a cam-operated, spring-loaded switch located in the distributor. On an eight-cylinder engine running at 3000rpm, these points must open and close 12,000 times per minute, which is 200 times per second.

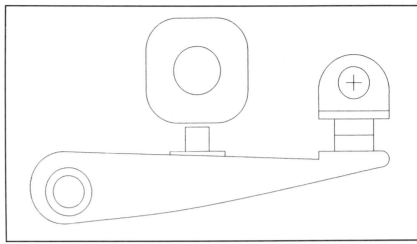

Most do-it-yourselfers are familiar with the point gap adjustment. When the points are held open at their widest, the gap should be a specified amount. Equally as important, be sure that when the points are closed, the contacts of the points are flush to one another.

rent flow.

The ballast resistor is located in the run circuit between the ignition switch and the coil. The bypass position of the ignition switch routes coil current around the ballast resistor while the engine is being cranked. This provides full coil power to the spark plugs while starting the engine. This is particularly important when trying to cold-start the engine.

If all this potential power can be created with full current flow to the ignition coil, why even have a ballast resistor? If the ballast resistor is eliminated from the system, arcing and excessive wear can occur at the points.

Condenser

The ignition condenser is an electrolytic capacitor. This device is a small metal can. Inside the can are two thin foil strips separated by a thin insulating material. The condenser acts like an electrical shock absorber to the primary ignition system. When the points open, the primary current attempts to keep flowing. Without the condenser, the current would continue to flow across the opening points. This arcing would slow the collapse of the magnetic field and limit the potential power of the secondary ignition.

The condenser is seldom the cause of driveability problems. In fact, during the heyday of the point/condenser system, I worked for several repair shops that routinely did not replace the condenser during a routine tune-up. Yet a defective condenser can cause poor idling and no-start misfiring.

The Points

The points are little more than a cam-operated switch. It is the opening and closing of the points in a point/condenser ignition system that control the current flow through the coil. As the crankshaft rotates, it drives the camshaft through gears or chains. A gear on the camshaft drives the shaft of the distributor. As the distributor shaft rotates, it turns a cam. This rotating cam opens and closes the

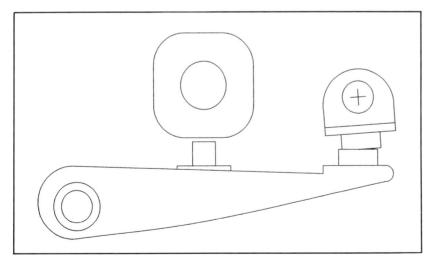

If the contacts of the points are not flush when the points are closed, uneven and rapid wear will occur. A set of properly installed points could last 15,000 to 20,000 miles of normal use. Poorly adjusted points may last less than 1,000 miles.

point. Therefore, the opening and closing of the points are synchronized to the rotation of the crankshaft and the movement of the pistons.

The points are an inadequate method of controlling the current for several reasons. First, as the points open there is a tendency for the current to continue flowing. This tendency can cause an arc and slow the speed of collapse of the magnetic field around the primary windings of the coil. As the speed of collapse slows, the potential output of the coil is decreased. In order to control this tendency to arc, the current flow through the primary is kept low. This also limits the potential output of the system. Secondly, the points become pitted and worn over time. This increases the resistance across the points and limits the current flow. Also, the rubbing block can wear, which will alter the timing. Finally, at extremely high engine rpms the points will tend to bounce open when they close.

Several things were done over the years to decrease the effects of these problems. Special metals were used at the contact points to decrease the wear and resistance factors. Holes were drilled in the contact points to reduce the possibility of arcing. But most of these only delayed the inevitable.

Since pitting and wear of the points can decrease the potential current flow through the primary and since the current flow in the primary can affect the output of the secondary, points need routine replacement. If the condenser is not perfectly matched to the rest of the primary ignition system, point pitting can be accelerated. Pitted points can cause erratic coil operation, which will result in misfire, stalling, and difficulty in starting. The ignition points need to be replaced at least every 15,000 miles to ensure the dependability of the ignition system.

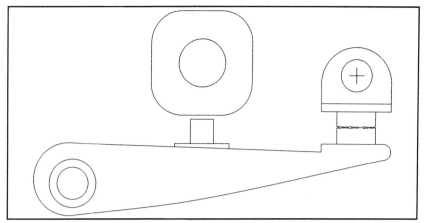

As the points are used, metal begins to transfer from one contact to the other. This results in pitting and poor contact.

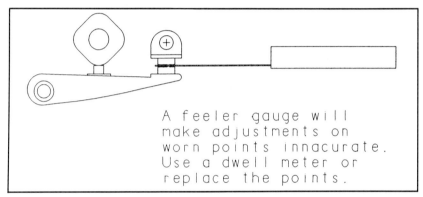

If a feeler gauge is used to check point gap that is pitted or has poor contact, the dwell will be incorrect. It used to be suggested that if you had to readjust the points, only use a dwell meter to do it. Actually, if the points are pitted badly enough to require the use of a dwell meter for adjustment, they are pitted badly enough to warrant replacement.

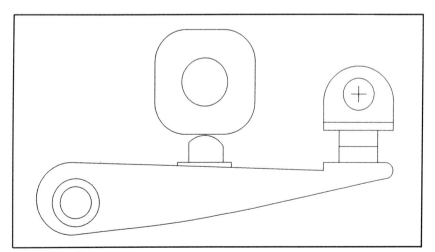

Another problem that can occur between tune-ups is a worn rubbing block. As the rubbing block wears, the ignition dwell changes. As the ignition dwell changes, the timing changes. A special lubricant called point lube is used on the rubbing block to decrease wear.

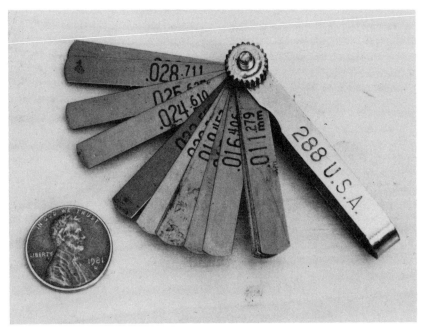

Feeler gauges come in all sizes and prices. These small feeler gauges are handy for getting into tight places like a distributor still mounted in the engine.

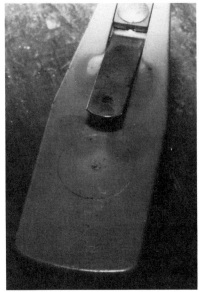

This rotor cost the company that owned the van $1,000. The van was towed in and the company's fleet mechanics worked on it for two days. Finally, they replaced the fuel injection computer. The van still did not start. The rotor had a pinhole under the electrode that was almost too small to see. The spark grounded to the distributor shaft and the engine would not start. The problem and $1,000 expense could have been avoided if the company had made it a policy to replace the rotor routinely.

Secondary Ignition Components
Coil

The ignition coil secondary consists of hundreds of windings of very thin wire. When current stops flowing through the primary, the magnetic field created by the primary current flow collapses. This collapsing magnetic field induces several thousand volts in the secondary. This is the voltage that is used to jump the gap of the spark plugs to fire the mixture in the cylinders.

Because of the relatively low current in this high-voltage side of the coil, there are relatively few problems in the secondary side of the coil. The problems that do occur are usually in the form of open circuits.

Coil Wire

The coil secondary output wire carries the high-voltage current from the coil to the distributor cap. The coil wire is normally 6 to 12in long and has a resistance of a few thousand ohms. This relatively high resistance helps to reduce the intensity of the radio signal created by the secondary ignition.

All arcing generates a radio signal. In addition to the arcing that occurs at the spark plug, there is also an arc inside the distributor cap between the cap and the rotor. All of the secondary ignition wiring has a high resistance to reduce the effect of this arcing.

The coil wire can suffer from several possible problems. As the coil wire ages its resistance tends to increase. Also, as the wire ages the insulating quality of the jacket decreases, which makes it possible for the high voltages being carried by the wire to penetrate the jacket and arc to ground. Corrosion can also affect the current carrying ability of the wire.

A defective coil wire can result in misfiring, no-start, and poor power.

Distributor Cap and Rotor

These two components operate as a team. The coil wire delivers the high voltage to the center terminal of the distributor cap. A carbon conductor carries the voltage to the center of the rotor. The rotor will have either a metal or carbon resistive conductor that carries the voltage to the tip of the rotor. The rotor mounts on the top of the distributor shaft and is driven by the camshaft. As the rotor rotates, it approaches either copper or aluminum conductors on the inside of the distributor cap and arcs to these conductors, which carry the voltage to the spark plug wires.

The distributor cap is prone to cracking, corrosion, and carbon tracking. Carbon tracking occurs when a microscopic crack or piece of dirt provides a current path to ground that is easier than the current path and plug wire. The rotor is subject to corrosion and perforation. Perforation occurs when the high voltage seeks and finds a ground through the rotor to the distributor shaft.

Routine replacement of the distributor cap and rotor can prevent unforeseen problems. It is not necessary to replace the cap and rotor at each tune-up as many professional technicians recommend, but they should be replaced at every other tune-up. As you read this, however, do not assume that the preceding statement means that your mechanic has been ripping you off for the past 10 years. There is no disservice in a mechanic charging you $30 or $40 extra at each tune-up to ensure that you have less chance of developing premature problems.

As discussed in Chapter 2, when replacing the distributor cap or rotor, I advise that you replace them as a set. I further advise that they be built by the same name brand manufacturer. I have had situations in the past where a mismatched cap and rotor caused the rotor air gap to be so large that the engine either failed to start or misfired.

Spark Plug Wires

Back in the late sixties and the early seventies, when I first got into the car repair business, we checked spark plug wires by starting the engine in a darkened shop. If there were sparks flying around under the hood it meant that one or more of the plug wires had perforated and was arcing to ground. For several years in the mid-seventies, I worked almost exclusively on fuel-injected imports. Upon returning to working on domestics in 1980, I was mystified at the poor quality of the plug wires. I remembered few problems in the seventies, now there were many problems.

The difference was not the

On many applications, the distributor cap can be pricey. Yet, routine replacement can reduce unnecessary problems.

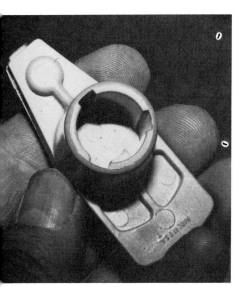

When deciding whether or not to replace a rotor, particularly if the problem is a no start, carefully inspect the inside of the rotor. Here, evidence of a hole through the rotor is grounding the spark.

Spark plug wires are an item that last longer today than they did in the sixties yet need to be replaced more often today than they did in the sixties. If this seems like double-talk, it is. Because today's engines run leaner, a plug wire that might have been adequate on a sixties engine would not do the job on a seventies engine.

At a glance, there is nothing wrong with this spark plug, yet the cylinder this plug was installed in had a 100-percent misfire. Looking at a spark plug to see if it is good is like sniffing the TV schedule to find the best programs. If there is any doubt about the condition of a spark plug, replace it.

The point/condenser ignition system requires accurate adjustments of the spark plug gap like any other ignition system. Considering the relatively weak spark and tendency for the tune-up specs to go out of adjustment, the gap adjustment is critical. These are just two of the many tools available to adjust spark plug gap.

quality of the wires, but rather the leaner air/fuel ratios demanded in the early eighties (and today). In the early seventies, we unwittingly compensated for defective spark plug wires by enriching the idle mixture screws on the carburetors. This option was not available on the sealed carburetors of the late seventies. Therefore, technicians were forced to replace defective spark plug wires rather than simply masking the problem with a richer mixture.

When replacing spark plug wires, the old adage, "You get what you pay for" is especially true. A $50 set of spark plug wires can easily outlast four $12 sets of wires. A marginal plug wire can cause a misfire when the engine is under an extreme load and on a modern fuel injected car can make the engine run rich.

Spark Plugs

Every tune-up includes replacing the spark plugs. There are many brands of spark plugs on the market, some good, some bad. Asking for opinions on which is the best brand of spark plug is like asking a group which is the best soft drink, it is largely a matter of personal opinion. What I have always done, when I had a choice, was to use the brand that the manufacturer installed at the factory. My thinking is that the manufacturer has a vested interest in choosing the spark plug that would provide the best driveability and has the least chance of requiring replacement within 12,000 miles. This method has rarely failed to provide either myself or my customers with good service.

The spark plug consists of a pair of electrodes separated by an air gap of between 0.028 and 0.075in. As the spark from the coil travels down the plug wire seeking ground, it must arc across this air gap. If this arc is exposed to a properly preheated, well atomized mixture of fuel

In addition to being used to adjust spark plug gap, this tool can also sense if the intensity of the spark traveling through the wire is sufficient to fire the spark plugs.

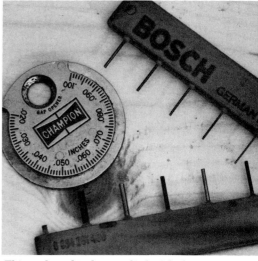

This tool used to be supplied with the tune-up kit for the OHV Volvos. It could gap the spark plugs, adjust the points gap, and even be used to adjust the valves.

and air, this spark will ignite the fuel.

As the spark plugs arc at a high frequency in high temperatures, the electrode material will slowly vaporize. This causes the gap to widen. The wider the gap, the higher the voltage required to initiate the spark across the gap. Eventually, the voltage required to initiate the spark across the gap will be greater than the coil is capable of generating and a misfire will occur.

Common problems associated with the spark plugs are misfiring and difficulty in starting.

Timing Control

The ignition timing must change as the engine is running to adjust to different engine speeds and loads. When the engine is running at an idle, the spark must begin at a point in crankshaft rotation that will allow for the spark to extinguish when the crankshaft is about 10deg after top dead center. Since the length of time the spark is jumping the gap is a relative constant, about 2.5 milliseconds, the spark must start sooner as the engine speed increases.

Centrifugal Advance

As an example, on a hypothetical engine the spark occurs at 10deg before top dead center when the engine is running at 1000rpm. At this speed, the spark extinguishes at about 10deg after top dead center. This means that the crankshaft has rotated 20deg since the initiation of spark. As the engine speed increases, the crankshaft rotates more degrees in the 2.5 milliseconds that the spark is jumping

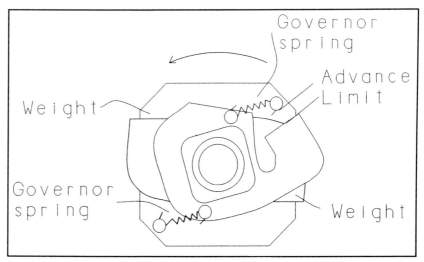

Timing on the point/condenser ignition system, like many electronic ignition systems, is controlled mechanically. The centrifugal advance system advances ignition timing as the speed of the engine increases. This is necessary because as the speed of the engine increases, the crankshaft turns further while the spark plug is firing. Since the goal is to end the burn at the same crankshaft position regardless of engine speed, the plug must be fired earlier at higher engine speeds.

the gap. At 2000rpm the crankshaft will rotate twice as much. If the timing at 1000rpm should be 10deg before TDC (top dead center), then the timing at 2000rpm should be about 30deg before TDC. As the engine speed continues to increase, the timing will need to continue to advance.

The amount of total advance, the upper limit of the advance, will vary depending on the design of the engine.

The change of timing in response to rpm is accomplished through a set of spring-loaded weights. As the speed of the engine increases, the weights swing out against spring tension. The cam that opens and closes the points, although it is mounted on the distributor shaft, is not part of the distributor shaft. The swinging weights cause the cam to rotate with respect to the distributor shaft. This advances the timing.

Vacuum Advance

At first glance this is a misnomer. The vacuum advance actually retards the timing when the engine is under a load. In most applications the vacuum advance is connected to ported vacuum. The advance unit receives no vacuum at an idle, but when the throttle is opened the vacuum advances the timing. As the load on the engine increases, the vacuum drops. As the vacuum drops, the timing is not advanced as much, it retards. Retarding the timing lowers the combustion temperature, therefore prevents detonation and decreases the potential of damage to the engine.

The job of the vacuum advance is to retard ignition timing when the engine vacuum drops. Engine vacuum drops when there is a load on the engine. Retarding the timing when the engine is loaded helps to prevent detonation or pinging.

Setting the Timing

There are almost as many different procedures to follow in the adjustment of initial timing in point/condenser ignition systems as there are different applications of point/condenser systems. The basic premise of adjusting or checking the initial timing is to disable the timing controls. Most point/condenser systems have a centrifugal advance system that advances the timing as the speed of the engine increases. A second timing control system is the vacuum advance that actually retards the timing when the engine is under a load.

For most applications, disconnect and plug the vacuum advance hose and lower the engine idle speed as low as possible. Under these conditions, any vacuum that might be in the vacuum hose cannot affect the timing and the engine speed will be too low for the centrifugal weight to advance the timing.

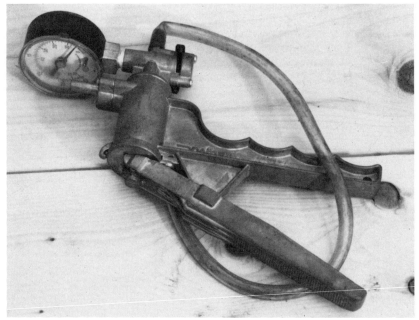

The vacuum advance uses manifold vacuum to advance the ignition timing when engine loads are low. The best way to test the diaphragm in the vacuum advance unit is with a hand-held vacuum pump.

This summary will usually work when the correct procedure for your application is not available. Be sure to follow the exact procedure for the application you are working on. This procedure can be found on the decal under the hood.

Troubleshooting
No Start

No start problems are probably the easiest to troubleshoot. There are three things required to get the engine started: fuel, air, and spark. In this book, I am confining myself to no-start problems related to the ignition system.

Confirm that there is air and fuel available to the engine. Disconnect one of the spark plug wires and perform the Ben Franklin test: Hand the plug wire to your nephew and crank the engine. If he begins to jump around, you know you have good spark. However, if you like your nephew or fear reprisal from his parents, insert a screwdriver into the plug wire and hold the screwdriver about a quarter inch from the engine block or cylinder head. Have someone crank the engine and observe the spark. If there is a good, crisp, blue spark, replace the spark plugs. If the spark is not a good, crisp, blue, move on to the next step.

Remove the coil wire from the distributor cap and hold it a quarter inch from ground. Crank the engine. If the spark is a good, crisp, blue, replace the distributor cap and rotor. Please keep in mind there is a possibility that the problem is a defective set of plug wires. The reason I do not suggest their replacement at this point is that it is unlikely all of the plug wires went completely bad at once. If replacing the distributor cap and rotor does not cure the problem, replace the plug wires.

If there is not a good spark from the coil wire, connect a test light to the negative terminal of the coil. Crank the engine. If the

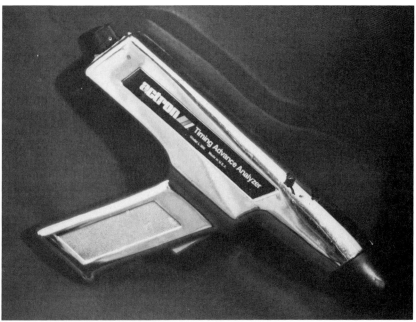

Timing lights can be frightfully expensive. The most essential features on a timing light are the brightness of the bulb and an inductive pickup, which clamps over the plug wire. A nice additional feature is a timing advance function that allows the technician to set the dial on the light to the timing specification and simply line the timing mark up to top dead center.

When troubleshooting a no-start condition, it is essential that three things be present, and that they be present at the right times and in the right amount. These three things are spark, fuel, and air. The easiest of these to check for is air.

39

test light blinks on and off as the engine is cranked, the problem is in the secondary side of the ignition system. Check the resistance of the coil wire. If the resistance is greater than 20,000 ohms, replace the coil wire. If the resistance is less than 20,000 ohms, replace the coil.

If the test light does not blink as the engine is cranked, the problem is in the primary. Move the test light to the positive terminal of the coil. Make sure the key is on. If the test light does not light, there is a problem in the power supply side of the primary ignition system. If the test light does light, then the problem is in the distributor, ground, side of the system.

If the problem was on the power side, crank the engine. If the test light illuminates while the engine is being cranked, check the resistance of the ballast resistor. The ballast resistor should have an extremely low resistance. Replace it if necessary. If the ballast resistor is good, repair the wiring in the bypass circuit. If the test light did not illuminate when the engine was cranked, check the circuit from the ignition switch up to where the current flow divides to go through the coil and the ballast resistor.

If the problem was on the distributor side, remove the distributor cap and inspect the points. If they appear to be pitted or burnt, replace the points and condenser. Check again for a pulse at the negative terminal of the coil with the test light. If the test light pulses when the engine is cranked, then the engine should now start. If the test light still does not pulse, check the distributor ground. If the distributor ground is good, repair the wire between the negative terminal of the coil and the distributor.

Starts But Does Not Continue to Run When the Key Is Released

The most common cause of this classic point/condenser symptom is a defective ballast resistor. However, any open in the main 12 volt power supply to the ignition coil can cause this symptom.

Misfire at Idle

Although generally an ignition misfire is in the secondary side of the ignition system, problems with the ignition points can cause symptoms that are exactly the same as defective spark plugs.

Before troubleshooting any misfire, it is essential to verify that the engine is in good condition. A compression test is a good starting point. If the valves are adjustable, be sure that they are properly adjusted.

With a pair of "sissy" pliers, remove and replace one plug wire at a time from the spark plugs. As each plug wire is removed, the engine rpm should drop. If one of the cylinders fails

To check for spark from the ignition coil to the spark plugs, remove one of the spark plug wires from a spark plug, insert a screwdriver in the end of the wire and hold it a quarter-inch from ground. Crank the engine. A spark should be present between the side of the screwdriver and ground.

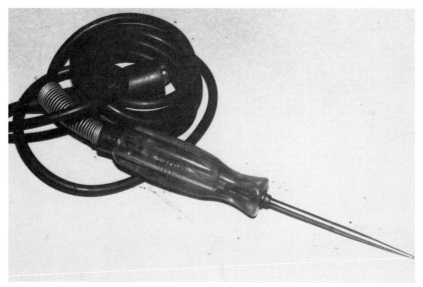

Use a test light on the negative terminal of the ignition coil to see if the points are opening and closing. With the engine cranking, the test light should be flickering on and off.

to produce as great a drop in rpm as the others, that cylinder is the source of the misfire.

Assuming the cylinder is in good condition and the valves are properly adjusted, remove the spark plug wire for that cylinder and check the resistance. The resistance should be less than 10,000 ohms per volt. If the resistance is correct, replace the spark plug. Unless the spark plugs are very new, replace them all. I was once in an auto parts store in Fort Worth, Texas, several years ago when an elderly cowboy came in and asked for a single spark plug. After brief jibing from the salesperson, the gentleman stated, "If you had a team of eight horses and one broke his leg you wouldn't shoot all eight would you?" This rhetorical question is humorous but does not hold up well in the real world of troubleshooting.

Misfire Under a Load

Assuming the engine is in good condition, begin troubleshooting this problem by checking the spark plug gap. If they are gapped properly, replace the spark plugs. Even new spark plugs can misfire under a load.

If replacing the spark plugs does not solve the problem, remove the distributor cap. Inspect the wiring to the points. Frayed wiring can cause an intermittent open circuit as the vacuum advance moves the breaker plate. The intermittent open can cause a misfire.

Lack of Power

There are many things that can cause a lack of power, some related to the ignition system, some not. Begin checking this problem by confirming the engine is in good condition as are the air and fuel filters.

If a lack of power is the result of problems in the ignition system, it is likely the problem is in the timing control system. To test the timing control system,

When trying to find a misfire at idle, one of the most effective tools is the engine analyzer. In the hands of an unskilled technician the machine is useless. In the hands of a skilled technician, however, the machine can precisely pinpoint a problem as being either fuel, spark, or compression related.

Many ignition related misfires are as simple as a plug wire that has fallen off. If the plug wire itself is in good condition, a little money can be saved by installing a new plug wire end.

connect a timing light to the engine. Disconnect the vacuum advance and plug in the hose. With the engine at idle speed, check the timing. Now raise the engine speed to 2000 to 2500rpm. If the timing does not advance, the centrifugal advance system is not working. Inspect the distributor weights. If they are free and move easily, replace the weight springs. If the springs are weak, they will allow the timing to advance all the way prematurely, even at idle. If the weights are frozen, use penetrating oil or whatever is necessary to free them. If they are badly corroded, it may be necessary to replace the distributor.

If, or when, the centrifugal advance is working properly, with the engine still at 2000 to 2500rpm, reconnect the vacuum hose to the vacuum advance. When the vacuum hose is reconnected, the timing should advance several degrees.

AMC/Jeep: Breakerless Inductive Discharge (BID)

4

1975-1977 American Motors/Jeep

We all have first memories about cars, first cars, first tickets, first... One of my memory "firsts" is my first car that had an electronic ignition system. It was a 1975 AMC Gremlin equipped with the BID ignition. My memories are not entirely fond. Within a few weeks of purchase, the car began a habit of intermittent shutdown on the freeway. At the time, I was like most other technicians, baffled by the mysteries of electronic ignition and feeling that in spite of twelve years of public school, a year of trade school, and three years toward a college degree, I would be incapable of learning how to repair my own car.

I took it to the dealer. The dealer threw some parts at it. I took it back to the dealer, he threw some more parts at it. I became resolved to the fact that someday, when the engine suddenly died on a busy freeway, I would become a hood ornament for a Peterbilt. Finally, after the car went out of warranty, I installed a point/condenser distributor out of a 1974 model. In my years of conducting seminars and workshops around the United States and the Pacific rim, I have discovered that I was not alone in these problems.

All of this would have been unnecessary if I had only known how simple this ignition system was.

Primary Ignition Components
Electronic Control Unit

The electronic control unit (ECU) controls current through the primary side of the ignition coil. There are five wires going into the ignition coil. The blue wire and the white wire are connected to the distributor pickup. A yellow wire is connected to the positive terminal of the coil. This wire provides power to the ECU. A green wire is connected to the negative terminal of the coil. With this wire, the computer controls the current flow through the coil. These four wires are grouped through a single connector. The fifth wire is connected to ground through a single wire connector.

The control unit sends a small alternating current through the blue and white wires to a sensor located in the distributor. This sensor alters the alternating current. As the ECU senses the change in the alternating current, it shuts off the current flow through the coil. As the current flow is shut off, the magnetic field it created collapses inducing a high-voltage spark in the secondary.

Sensor coil

The sensor coil is located in the distributor. An alternating current is sent from the ECU through the sensor coil. Press fit on the distributor shaft is a trigger wheel. As the trigger wheel rotates, it passes near the core of the sensor coil and affects the flow of the current through the coil. The ECU reacts to this rip-

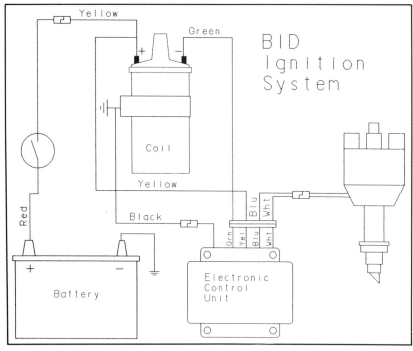

The BID ignition system was among the first electronic ignition systems to make it into mass production. While the system usually worked quite well, it did have a tendency to shut down occasionally. This problem, though annoying, was easily repaired by removing the four wire connectors and soldering the wires together.

ple in the alternating current through the sensor and shuts off the primary current flow creating the spark.

Ignition Switch

The ignition switch controls whether current will be available to the coil and ignition module. Unlike the point/condenser ignition system, the BID system uses the same current path when starting the engine that it uses when the engine is running. There is no ballast resistor in the BID system.

Secondary Ignition Components
Coil

The ignition coil secondary consists of hundreds of windings of very thin wire. When current stops flowing through the primary, the magnetic field created by the primary current flow collapses. This collapsing magnetic field induces several thousand volts in the secondary. This is the voltage that is used to jump the gap of the spark plugs to fire the mixture in the cylinders.

Because of the relatively low current in this high-voltage side of the coil, there are relatively few problems in the secondary side of the coil. The problems that do occur are usually in the form of open circuits.

Coil Wire

The coil secondary output wire carries the high-voltage current from the coil to the distributor cap. The coil wire is normally 6 to 12in long and has a resistance of a few thousand ohms. This relatively high resistance helps to reduce the intensity of the radio signal created by the secondary ignition.

All arcing generates a radio signal. In addition to the arcing that occurs at the spark plug, there is also an arc inside the distributor cap between the cap and the rotor. All of the secondary ignition wiring has a high resistance to reduce the effect of this arcing.

The coil wire can suffer from several possible problems. As the coil wire ages its resistance tends to increase. Also, as the wire ages the insulating quality of the jacket decreases, which makes it possible for the high voltages being carried by the wire to penetrate the jacket and arc to ground. Corrosion can also affect the current carrying ability of the wire.

A defective coil wire can result in misfiring, no-start, and poor power.

The Distributor Cap and Rotor

These two components operate as a team. The coil wire delivers the high voltage to the center terminal of the distributor cap. A carbon conductor carries the voltage to the center of the rotor. The rotor will have either a metal or carbon resistive conductor that carries the voltage to the tip of the rotor. The rotor mounts on the top of the distributor shaft and is driven by the camshaft. As the rotor rotates, it approaches either copper or aluminum conductors on the inside of the distributor cap and arcs to these conductors, which carry the voltage to the spark plug wires.

The distributor cap is prone to cracking, corrosion, and carbon tracking. Carbon tracking occurs when a microscopic crack or piece of dirt provides a current path to ground that is easier than the current path and plug wire. The rotor is subject to corrosion and perforation. Perforation occurs when the high voltage seeks and finds a ground through the rotor to the distributor shaft.

Routine replacement of the distributor cap and rotor can prevent unforeseen problems. It is not necessary to replace the cap and rotor at each tune-up as many professional technicians recommend, but they should be replaced at every other tune-up.

As you read this, however, do not assume that the preceding statement means that your mechanic has been ripping you off for the past 10 years. There is no disservice in a mechanic charging you $30 or $40 extra at each tune-up to ensure that you have less chance of developing premature problems.

As discussed earlier, when replacing the distributor cap or rotor, I advise that you replace them as a set. I further advise that they be built by the same name brand manufacturer. I have had situations in the past where a mismatched cap and rotor caused the rotor air gap to be so large that the engine either failed to start or misfired.

Spark Plug Wires

Back in the late sixties and the early seventies, when I first got into the car repair business, we checked spark plug wires by starting the engine in a darkened shop. If there were sparks flying around under the hood it meant that one or more of the plug wires had perforated and was arcing to ground. For several years in the mid-seventies, I worked almost exclusively on fuel injected imports. Upon returning to working on domestics in 1980, I was mystified at the poor quality of the plug wires. I remembered few problems in the seventies, now there were many problems.

The difference was not the quality of the wires, but rather the leaner air/fuel ratios demanded in the early eighties (and today). In the early seventies, we unwittingly compensated for defective spark plug wires by enriching the idle mixture screws on the carburetors. This option was not available on the sealed carburetors of the late seventies. Therefore, technicians were forced to replace defective spark plug wires rather than simply masking the problem with a richer mixture.

When replacing spark plug

wires, the old adage, "You get what you pay for" is especially true. A $50 set of spark plug wires can easily outlast four $12 sets of wires. A marginal plug wire can cause a misfire when the engine is under an extreme load and on a modern fuel injected car can make the engine run rich.

Spark Plugs

Every tune-up includes replacing the spark plugs. There are many brands of spark plugs on the market, some good, some bad. Asking for opinions on which is the best brand of spark plug is like asking a group which is the best soft drink, it is largely a matter of personal opinion. What I have always done, when I had a choice, was to use the brand that the manufacturer installed at the factory. My thinking is that the manufacturer has a vested interest in choosing the spark plug that would provide the best driveability and has the least chance of requiring replacement within 12,000 miles. This method has rarely failed to provide either myself or my customers with good service.

The spark plug consists of a pair of electrodes separated by an air gap of between 0.028 and 0.075in. As the spark from the coil travels down the plug wire seeking ground, it must arc across this air gap. If this arc is exposed to a properly preheated, well atomized mixture of fuel and air, this spark will ignite the fuel.

As the spark plugs arc at a high frequency in high temperatures, the electrode material will slowly vaporize. This causes the gap to widen. The wider the gap, the higher the voltage required to initiate the spark across the gap. Eventually, the voltage required to initiate the spark across the gap will be greater than the coil is capable of generating and a misfire will occur.

Common problems associated with the spark plugs are misfiring and difficulty in starting.

Timing Control

The ignition timing must change as the engine is running to adjust to different engine speeds and loads. When the engine is running at an idle, the spark must begin at a point in crankshaft rotation that will allow for the spark to extinguish when the crankshaft is about 10deg after top dead center. Since the length of time the spark is jumping the gap is a relative constant, about 2.5 milliseconds, the spark must start sooner as the engine speed increases.

Centrifugal Advance

As an example, on a hypothetical engine the spark occurs at 10deg before top dead center when the engine is running at 1000rpm. At this speed, the spark extinguishes at about 10deg after top dead center. This means that the crankshaft has rotated 20deg since the initiation of spark. As the engine speed increases, the crankshaft rotates more degrees in the 2.5 milliseconds that the spark is jumping the gap. At 2000rpm the crankshaft will rotate twice as much. If the timing at 1000rpm should be 10deg before TDC (top dead center), then the timing at 2000rpm should be about 30deg before TDC. As the engine speed continues to increase, the timing will need to continue to advance. The amount of total advance, the upper limit of the advance, will vary depending on the design of the engine.

The change of timing in response to rpm is accomplished through a set of spring-loaded weights. As the speed of the engine increases, the weights swing out against spring tension. The trigger wheel that trips the sensor, although it is mounted on the distributor shaft, is not part of the distributor shaft. The swinging weights cause the trigger wheel to rotate with respect to the distributor shaft. This advances the timing.

Vacuum Advance

At first glance this is a mis-

In order to get maximum power from the engine, combustion must end at a certain point (about 10deg) after top dead center. To achieve this, the ignition timing must be advanced as the engine speed increases.

nomer. The vacuum advance actually retards the timing when the engine is under a load. In most applications the vacuum advance is connected to ported vacuum. The advance unit receives no vacuum at an idle, but when the throttle is opened the vacuum advances the timing. As the load on the engine increases, the vacuum drops. As the vacuum drops, the timing is not advanced as much, it retards. Retarding the timing lowers the combustion temperature, and therefore prevents detonation and decreases the potential of damage to the engine.

Setting the Timing

There are almost as many different procedures to follow in the adjustment of initial timing in BID ignition systems as there are different applications of BID systems. The basic premise of adjusting or checking the initial timing is to disable the timing controls. Most BID systems have a centrifugal advance system which advances the timing as the speed of the engine increases. A second timing control system is the vacuum advance, which actually retards the timing when the engine is under a load.

For most applications, disconnect and plug the vacuum advance hose and lower the engine idle speed as low as possible. Under these conditions, any vacuum that might be in the vacuum hose cannot affect the timing and the engine speed will be too low for the centrifugal weight to advance the timing.

This summary will usually work when the correct procedure for your application is not available. Be sure to follow the exact procedure for the application you are working on. This procedure can be found on the decal under the hood.

Troubleshooting
No start

No-start problems are probably the easiest to troubleshoot. There are three things required to get the engine started: fuel, air, and spark. In this book, I am confining myself to no-start problems related to the ignition system.

Confirm that there is air and fuel available to the engine. Disconnect one of the spark plug wires and perform the Ben Franklin test: Hand the plug wire to your nephew and crank the engine. If he begins to jump around, you know you have good spark. However, if you like your nephew or fear reprisal from his parents insert a screwdriver into the plug wire and hold the screwdriver about a quarter inch from the engine block or cylinder head. Have someone crank the

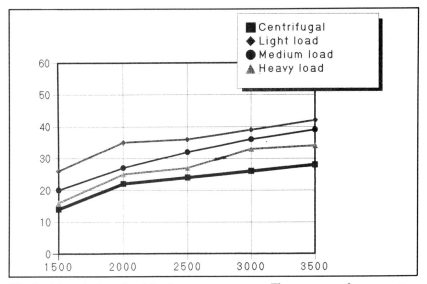

The ignition timing should advance as the engine speed increases. But as the engine load increases, the tendency for detonation to occur also increases. The vacuum advance system must reduce timing advance as the engine vacuum decreases while the engine is under a load.

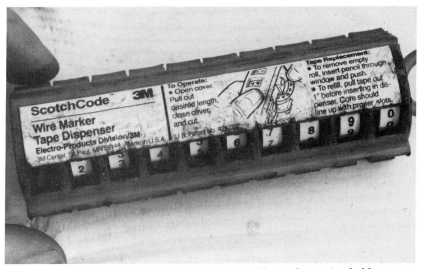

Many ignition misfires occur right after a tune-up. Even the most skilled technician occasionally crosses spark plug wires. These labeling tapes are available in electronics hobby stores and will help keep the plug wires identified while the tune-up is being done.

engine and observe the spark. If there is a good, crisp, blue spark, replace the spark plugs. If the spark is not a good, crisp, blue, move on to the next step.

Remove the coil wire from the distributor cap and hold it a quarter inch from ground. Crank the engine. If the spark is a good, crisp, blue, replace the distributor cap and rotor. Please keep in mind there is a possibility that the problem is a defective set of plug wires. The reason I do not suggest their replacement at this point is that it is unlikely all of the plug wires went completely bad at once. If replacing the distributor cap and rotor does not cure the problem, replace the plug wires.

If there is not a good spark from the coil wire, connect a test light to the negative terminal of the coil. Crank the engine. If the test light blinks on and off as the engine is cranked, the problem is in the secondary side of the ignition system. Check the resistance of the coil wire. If the resistance is greater than 20,000 ohms, replace the coil wire. If the resistance is less than 20,000 ohms, replace the coil.

If the test light does not blink as the engine is cranked, the problem is in the primary. Move the test light to the positive terminal of the coil. Make sure the key is on. If the test light does not light, there is a problem in the power supply side of the primary ignition system. If the test light does light, the problem is on the module side of the coil.

Testing the Distributor Pickup

Testing the distributor pickup used in the BID system may be the most difficult of all the distributor pickups used in electronic ignition systems. When testing any component, I like to use a method that actually tests the presence and quality of the signal it produces. Since the BID distributor pickup must receive an alternating current from the module and since the affect of the pickup on the signal is minimal, there is no direct way to test the signal. A less than thorough method is to check the resistance of the distributor pickup.

Disconnect the distributor pickup from the wiring harness between the pickup and the ignition module. Connect an ohmmeter to the two wires to the pickup. If the pickup is good, the resistance should be 1.6 to 2.4 ohms. If the resistance is within these specifications, leave the ohmmeter connected and massage the sensor plug while watching the ohmmeter. Any fluctuation in the resistance indicates deterioration of the insulation in the plug. Replace the sensor if necessary. It was this problem that inspired me to replace the electronic ignition distributor in my Gremlin with a point/condenser distributor.

If the sensor passes all the resistance tests, check the clearance between the trigger wheel and the sensor base. This clearance must be 0.050in (1.27mm). If the clearance is good, replace the pickup.

Testing the Module

If the engine still does not start after the pickup is replaced, replace the module. If you are like me, that last statement is probably irritating. Unfortunately, there is no better method of troubleshooting the ignition module than by process of elimination.

Starts But Does Not Continue to Run When the Key Is Released

Since the BID system does not use a ballast resistor, it is unlikely that the cause of this symptom is related to the ignition system. Connect a test light to the positive terminal of the ignition coil. Crank the engine while watching the test light. If the test light is lit while the engine is being cranked but goes out when the key is released, replace the ignition switch.

If the test light does not go out when the key is released, the problem is most likely not related to the ignition system. Check the carburetor. Also check the intake system for evidence of vacuum leaks.

Dies While Driving Down the Road

This is a fairly common and annoying problem with this system. Almost always it is a result of faulty connections. Check the harness connections of the ignition module for corrosion or evidence of deterioration in the insulation of the connector. If the connector and connections look good, check the ignition pickup and module.

Disconnect the distributor pickup from the wiring harness between the pickup and the ignition module. Connect an ohmmeter to the two wires of the pickup. Tap on the sensor lightly with the butt end of a screwdriver. Heat the sensor with a light and repeat the tap test while the sensor is still warm. If the pickup is good, the resistance should remain at 1.6 to 2.4 ohms. If the resistance is within these specifications, leave the ohmmeter connected and massage the sensor plug while watching the ohmmeter. Any fluctuation in the resistance indicates deterioration of the insulation in the plug. Replace the sensor if necessary.

If the sensor passes all the resistance tests, check the clearance between the trigger wheel and the sensor base. This clearance must be 0.050in (1.27mm). If the clearance is good, replace the pickup.

Testing the Module

If the engine still dies at inconvenient times after the pickup is replaced, replace the module. There is no better method of troubleshooting the ignition module than by process of elimination.

Misfire at Idle

This is almost always never the fault of the BID system. This symptom is almost always the fault of some component in the secondary side of the ignition system.

Before troubleshooting any misfire, it is essential to verify that the engine is in good condition. A compression test is a good starting point. If the valves are adjustable, be sure that they are properly adjusted.

With a pair of "sissy" pliers remove and replace one plug wire at a time from the spark plugs. As each plug wire is removed, the engine rpm should drop. If one of the cylinders fails to produce as great a drop in rpm as the others, that cylinder is the source of the misfire.

Assuming the cylinder is in good condition and the valves are properly adjusted, remove the spark plug wire for that cylinder and check the resistance. The resistance should be less than 10,000 ohms per volt. If the resistance is correct, replace the spark plug. Unless the spark plugs are very new, replace them all.

Misfire Under a Load

Assuming the engine is in good condition, begin troubleshooting this problem by checking the spark plug gap. If they are gapped properly, replace the spark plugs. Even new spark plugs can misfire under a load.

If replacing the spark plugs does not solve the problem, remove the distributor cap. Inspect the wiring to the points. Frayed wiring can cause an intermittent open circuit as the vacuum advance moves the breaker plate. The intermittent open can cause a misfire.

Lack of Power

There are many things that can cause a lack of power, some related to the ignition system, some not. Begin checking this problem by confirming the engine is in good condition as are the air and fuel filters.

If a lack of power is the result of problems in the ignition system, it is likely the problem is in the timing control system. To test the timing control system, connect a timing light to the engine. Disconnect the vacuum advance and plug in the hose. With the engine at idle speed, check the timing. Now raise the engine speed to 2000 to 2500rpm. If the timing does not advance, the centrifugal advance system is not working. Inspect the distributor weights. If they are free and move easily, replace the weight springs. If the springs are weak, they will allow the timing to advance all the way prematurely, even at idle. If the weights are frozen, use penetrating oil or whatever is necessary to free them. If they are badly corroded, it may be necessary to replace the distributor.

If, or when, the centrifugal advance is working properly, with the engine still at 2000 to 2500rpm, reconnect the vacuum hose to the vacuum advance. When the vacuum hose is reconnected, the timing should advance several degrees.

Solid State Ignition (SSI)

5

In 1978, AMC adopted the Solid State Ignition (SSI) system and used it until 1987. Applications that have used the SSI system 1978-1979 V-8 AMC applications are the 1978-1984 six cylinder and V-8 Jeep applications and the AMC/Jeep 1984-1987 2.5 liter four cylinder applications. This system is essentially the same as the Ford SSI system.

Primary Ignition Components
Electronic Control Unit

There are six wires connected to the ECU on all except the 1982 through 1987 products. The 1982-1987 models might have two additional wires connected to the engine control computer. Through these two wires the Engine Control Module (ECM) controls the ignition timing. The ECU replaces the points and ballast resistor of the conventional point/condenser ignition system.

In the six wires, there are two and four wire connectors. The two wire connector will have a white or light blue wire. This wire will be connected to either the ignition switch or the starter solenoid. When the ignition switch is in the crank position, the ECU receives voltage from this wire. This allows full voltage to module when the engine is being started.

The second wire of the two wire connector will be red or red/white. This wire carries full system voltage to the ECU while the engine is running.

The four wire connector has a black, violet, orange, and dark green wire. The orange and violet wires are connected to a reluctance pickup in the distributor. The black wire is a ground, while a dark green wire is connected to the negative terminal of the coil.

As the ECU receives pulses from the distributor pickup, it switches the ground for the ignition coil on and off. When the ground for the ignition coil is switched off, the magnetic field built by the current flow through the primary collapses. As the magnetic field collapses, it induces a high voltage in the secondary.

For the models that have the extra two wire connector, a yellow wire runs from the ECU to the ECM. This wire carries a signal from the ECM to the ECU to control the ignition timing. The extra wire out of the ECU goes only as far as the connector.

Distributor pickup

The distributor pickup is a variable reluctance transducer. This sensor produces an alternating current as the distributor rotates. The sensor consists of a coil of wire, permanent magnet, and reluctor wheel. The reluctor wheel is part of the distributor shaft. As the reluctor wheel rotates, it distorts the magnetic field back and forth across the coil of wire. This distorting induces an alternating current in the coil of wire. As the speed of distributor shaft rotation increases, the amplitude (voltage output) of the sensor increases. More important, since the AC

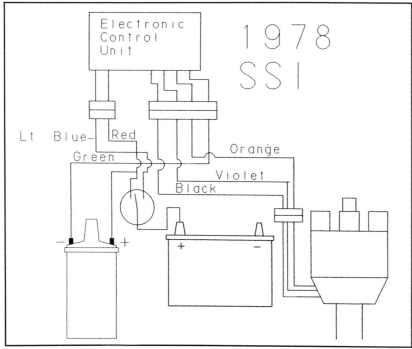

The SSI system was introduced by AMC in the late seventies. Designed by Ford, it used many components that were identical to the Ford applications. AMC used the SSI system until AMC was bought out by Chrysler.

signal is a sine wave, the frequency of the sine wave changes as the speed of the distributor shaft changes. The frequency of the sine wave tells the ignition module the speed and position of the distributor shaft. Since the distributor shaft is synchronized to the cam shaft, the ignition is synchronized to the mechanical components of the engine. The electronic control unit uses the signal from the pickup coil as a trigger source for the firing of the coil.

Ignition Switch

The ignition switch controls whether current will be available to the coil and ignition module. Unlike the point/condenser ignition system, the BID system uses the same current path when starting the engine that it uses when the engine is running. There is no ballast resistor in the SSI system.

Secondary Ignition Components
Coil

The ignition coil secondary consists of hundreds of windings of very thin wire. When current stops flowing through the primary, the magnetic field created by the primary current flow collapses. This collapsing magnetic field induces several thousand volts in the secondary. This is the voltage that is used to jump the gap of the spark plugs to fire the mixture in the cylinders.

Because of the relatively low current in this high-voltage side of the coil, there are relatively few problems in the secondary side of the coil. The problems that do occur are usually in the form of open circuits.

Coil Wire

The coil secondary output wire carries the high-voltage current from the coil to the distributor cap. The coil wire is normally 6 to 12in long and has a resistance of a few thousand ohms. This relatively high resistance helps to reduce the intensity of the radio signal created by the secondary ignition.

All arcing generates a radio signal. In addition to the arcing that occurs at the spark plug, there is also an arc inside the distributor cap between the cap and the rotor. All of the secondary ignition wiring has a high resistance to reduce the affect of this arcing.

The coil wire can suffer from several possible problems. As the coil wire ages its resistance tends to increase. Also, as the wire ages the insulating quality of the jacket decreases, which makes it possible for the high voltages being carried by the wire to penetrate the jacket and arc to ground. Corrosion can also affect the current carrying ability of the wire.

A defective coil wire can result in misfiring, no-start, and poor power.

The Distributor Cap and Rotor

These two components operate as a team. The coil wire delivers the high voltage to the center terminal of the distributor cap. A carbon conductor carries the voltage to the center of the rotor. The rotor will have either a metal or carbon resistive conductor that carries the voltage to the tip of the rotor. The rotor mounts on the top of the distributor shaft and is driven by the camshaft. As the rotor rotates, it approaches either copper or aluminum conductors on the inside of the distributor cap and arcs to these conductors which carry the voltage to the spark plug wires.

The distributor cap is prone to cracking, corrosion, and carbon tracking. Carbon tracking occurs when a microscopic crack or piece of dirt provides a current path to ground that is easier than the current path and plug wire. The rotor is subject to corrosion and perforation. Perforation occurs when the high voltage seeks and finds a ground through the rotor to the distributor shaft.

Routine replacement of the distributor cap and rotor can prevent unforeseen problems. It is not necessary to replace the cap and rotor at each tune-up as many professional technicians recommend, but they should be replaced at every other tune-up. As you read this, however, do not assume that the preceding statement means that your mechanic has been ripping you off for the past 10 years. There is no disservice in a mechanic charging you $30 or $40 extra at each tune-up to ensure that you have less chance of developing premature problems.

As discussed in Chapter 2, when replacing the distributor cap or rotor, I advise that you replace them as a set. I further advise that they be built by the same name brand manufacturer. I have had situations in the past where a mismatched cap and rotor caused the rotor air gap to be so large that the engine either failed to start or misfired.

Spark Plug Wires

Back in the late sixties and the early seventies, when I first got into the car repair business, we checked spark plug wires by starting the engine in a darkened shop. If there were sparks flying around under the hood it meant that one or more of the plug wires had perforated and was arcing to ground. For several years in the mid-seventies, I worked almost exclusively on fuel injected imports. Upon returning to working on domestics in 1980, I was mystified at the poor quality of the plug wires. I remembered few problems in the seventies, now there were many problems.

The difference was not the quality of the wires, but rather the leaner air/fuel ratios demanded in the early eighties (and today). In the early seventies, we unwittingly compensated for defective spark plug wires by enriching the idle mixture screws on the carburetors. This option was not available on the sealed carburetors of the late seventies. Therefore, technicians were forced to replace defective spark plug wires rather than simply masking the problem with a richer mixture.

When replacing spark plug wires, the old adage, "You get what you pay for" is especially true. A $50 set of spark plug wires can easily outlast four $12 sets of wires. A marginal plug wire can cause a misfire when the engine is under an extreme load and on a modern fuel injected car can make the engine run rich.

Spark Plugs

Every tune-up includes replacing the spark plugs. There are many brands of spark plugs on the market, some good, some bad. Asking for opinions on which is the best brand of spark plug is like asking a group which is the best soft drink, it is largely a matter of personal opinion. What I have always done, when I had a choice, was to use the brand that the manufacturer installed at the factory. My thinking is that the manufacturer has a vested interest in choosing the spark plug that would provide the best driveability and has the least chance of requiring replacement within 12,000 miles. This method has rarely failed to provide either myself or my customers with good service.

The spark plug consists of a pair of electrodes separated by an air gap of between 0.028 and 0.075in. As the spark from the coil travels down the plug wire seeking ground, it must arc across this air gap. If this arc is exposed to a properly preheated, well atomized mixture of fuel and air, this spark will ignite the fuel.

As the spark plugs arc at a high frequency in high temperatures, the electrode material will slowly vaporize. This causes the gap to widen. The wider the gap, the higher the voltage required to initiate the spark across the gap. Eventually, the voltage required to initiate the spark across the gap will be greater than the coil is capable of generating and a misfire will occur.

Common problems associated with the spark plugs are misfiring and difficulty in starting.

Timing Control

The ignition timing must change as the engine is running to adjust to different engine speeds and loads. When the engine is running at an idle, the spark must begin at a point in crankshaft rotation that will allow for the spark to extinguish when the crankshaft is about 10deg after top dead center. Since the length of time the spark is jumping the gap is a relative constant, about 2.5 milliseconds, the spark must start sooner as the engine speed increases.

Centrifugal Advance

As an example, on a hypothetical engine the spark occurs at 10deg before top dead center when the engine is running at 1000rpm. At this speed, the spark extinguishes at about 10deg after top dead center. This means that the crankshaft has rotated 20deg since the initiation of spark. As the engine speed increases, the crankshaft rotates more degrees in the 2.5 milliseconds that the spark is jumping the gap. At 2000rpm the crankshaft will rotate twice as much. If the timing at 1000rpm should be 10deg before TDC (top dead center), then the timing at 2000rpm should be about 30deg before TDC. As the engine speed continues to increase, the timing will need to continue to advance. The amount of total advance, the upper limit of the advance, will vary depending on the design of the engine.

The change of timing in response to RPM is accomplished through a set of spring-loaded weights. As the speed of the engine increases, the weights swing out against spring tension. The trigger wheel that trips the sensor, although it is mounted on the distributor shaft, is not part of the distributor shaft. The swinging weights cause the trigger wheel to rotate with respect to the distributor shaft. This advances the timing.

Vacuum Advance

At first glance this is a misnomer. The vacuum advance actually retards the timing when the engine is under a load. In most applications the vacuum advance is connected to ported vacuum. The advance unit receives no vacuum at an idle, but when the throttle is opened the vacuum advances the timing. As the load on the engine increases, the vacuum drops. As the vacuum drops, the timing is not advanced as much, it retards. Retarding the timing lowers the combustion temperature, therefore prevents detonation and decreases the potential of damage to the engine.

Setting the Timing

When setting the ignition timing, it is best to follow the instructions outlined on the EPA sticker under the hood. If the instructions are missing or obliterated, disconnect the vacuum advance and adjust to the specification listed in this book.

Troubleshooting
No Start

Like most ignition systems, a no-start condition in the SSI ignition system is among the easiest problems to troubleshoot. There are three things required to get the engine started. These things are air, fuel, and spark. Since the topic of this book is ignition systems, I will concentrate on the ignition problems that can prevent the engine from starting.

Begin testing the ignition

system by disconnecting a plug wire from one of the spark plugs. Stick a plastic-handled screwdriver in the end of one of the plug wires. While holding the screwdriver near ground, have someone crank the engine. If there is no spark, remove the coil wire from the distributor cap and place the screwdriver in the end of the coil wire. Hold the screwdriver a quarter inch from ground. Again, have someone crank the engine. If there is a spark, replace the distributor rotor and cap.

If there is no spark, connect a test light to the negative terminal of the coil. While you watch the test light, have someone crank the engine. If the test light blinks, the secondary windings of the ignition coil are defective and the coil should be replaced. To verify that the secondary windings are defective, use an ohmmeter to check the resistance of the secondary. The resistance at 75deg F should be between 7,700 and 9,300 ohms. If it does not blink, there is a problem in the primary side of the ignition system.

Checking the pickup coil

Begin troubleshooting the primary side of the ignition system by checking the pickup coil. There are several viable ways to test the pickup coil, some are more thorough than others. One of the most commonly suggested methods of testing the pickup coil is to use an ohmmeter. The resistance between the orange and violet wire through the pickup coil should be between 400 and 800 ohms. If the resistance through the pickup coil is considerably different from these readings, replace it. If the resistance is less than 100 ohms outside of these specifications, the no-start problem is likely to have another cause.

The ohmmeter method of testing the pickup coil is incomplete. While it tests the condition of the coil of wire, which is the most likely part of the pickup coil to be defective, it does not test the condition of the permanent magnet or the air gap between the pickup and the reluctor wheel. A better test is to disconnect the connector near the distributor or ignition module that contains the orange and violet wires. Connect an AC voltmeter between these wires on the distributor side of the connector. Have someone crank the engine while you watch the voltmeter. If the pickup generates less than 0.5 volt, replace the pickup. In other ignition systems, it would be important to check the air gap between the pickup and the reluctor wheel. However, in the SSI ignition this air gap is nonadjustable. If the air gap is incorrect it would require replacing the pickup coil.

Testing the Module

There is no convenient way to test the ignition module directly. Therefore, it is best tested through a process of elimination. If a test light connected to the negative side of the coil does not blink when the engine is cranked, and if the pickup tests good, replace the module.

Starts But Does Not Continue to Run When the Key Is Released

Like a point/condenser ignition system, the BID system uses a ballast resistor wire to limit current flow through the primary when the engine is running. A bypass circuit allows full current flow when the engine is being cranked.

Connect a test light to the positive terminal of the coil and turn the ignition switch to the run position. If the light is not on, crank the engine. If the light is on while the engine is being cranked but goes out when the key is released, test the ballast resistor wire. The ballast resistor wire should have between 1.3 and 1.4 ohms of resistance. If the resistance is outside of this specification, replace the ballast resistor.

If the ballast resistor wire is good, or if replacing the ballast resistor did not solve the problem, inspect the wiring from the crank position of the ignition switch to the ignition terminal of the starter solenoid. If the ballast wiring is good, or if the repairs do not cure the symptom, replace the ignition module.

Dies While Driving Down the Road

In my experience with this system, an intermittent shutdown is one of the more common problems. By observation, the severity of the problem and the frequency of the problem seemed to be heat related.

Although I caution the reader that the pickup coil could be at fault and suggest that an AC voltage output test be done, the symptom is almost always caused by the module. What I suggest to professional technicians is that they order the ignition module, get it on its way, then test the pickup coil. If the pickup coil is found to be defective, most parts houses will accept an unopened box with an electrical or electronic component.

Misfire at Idle

This is almost always never the fault of the SSI system and almost always the fault of some component in the secondary side of the ignition system.

Before troubleshooting any misfire, it is essential to verify that the engine is in good condition. A compression test is a good starting point. If the valves are adjustable, be sure that they are properly adjusted.

With a pair of "sissy" pliers remove and replace one plug wire at a time from the spark plugs. As each plug wire is removed, the engine rpm should drop. If one of the cylinders fails to produce as great a drop in rpm as the others, that cylinder is the source of the misfire.

Assuming the cylinder is in good condition and the valves are properly adjusted, remove the spark plug wire for that cylinder and check the resistance. The resistance should be less than 10,000 ohms per volt. If the resistance is correct, replace the spark plug. Unless the spark plugs are very new, replace them all.

Misfire Under a Load

Assuming the engine is in good condition, begin troubleshooting this problem by checking the spark plug gap. If they are gapped properly, replace the spark plugs. Even new spark plugs can misfire under a load.

If replacing the spark plugs does not solve the problem, remove the distributor cap. Inspect the wiring to the points. Frayed wiring can cause an intermittent open circuit as the vacuum advance moves the breaker plate. The intermittent open can cause a misfire.

Lack of Power

There are many things that can cause a lack of power, some related to the ignition system, some not. Begin checking this problem by confirming the engine is in good condition as are the air and fuel filters.

If a lack of power is the result of problems in the ignition system, it is likely the problem is in the timing control system. To test the timing control system, connect a timing light to the engine. Disconnect the vacuum advance and plug in the hose. With the engine at idle speed, check the timing. Now raise the engine speed to 2000 to 2500rpm. If the timing does not advance, the centrifugal advance system is not working. Inspect the distributor weights. If they are free and move easily, replace the weight springs. If the springs are weak, they will allow the timing to advance all the way prematurely, even at idle. If the weights are frozen, use penetrating oil or whatever is necessary to free them. If they are badly corroded, it may be necessary to replace the distributor.

If, or when, the centrifugal advance is working properly, with the engine still at 2000 to 2500rpm, reconnect the vacuum hose to the vacuum advance. When the vacuum hose is reconnected, the timing should advance several degrees.

AMC/Renault Ducellier Electronic Ignition

6

From 1983 through 1987, AMC/Renault used the Ducellier electronic ignition system. Among the systems used on domestic applications, it is unique because it incorporates full timing control within the Ignition Control Module (ICM).

Located in the flywheel housing of the engine is the Top Dead Center (TDC) sensor. This is the Ducellier system equivalent to the distributor pickup. As the flywheel rotates, the TDC sensor generates a sine wave, the frequency of which is directly proportional to the speed of the engine. This sine wave signal is fed into the ICM by way of two of its terminals in its three-terminal connector. This signal not only signals crankshaft position but also engine speed. The ICM uses the engine speed information much the same way a standard ignition system advances the timing through centrifugal weights. The higher the engine speed, the more the ICM advances the timing.

The Ducellier ignition system has a vacuum sensor located on the ICM. Connected to the intake manifold through a vacuum hose, the ICM uses manifold vacuum information in the same way the vacuum advance unit of a standard ignition system uses that information. A drop in the manifold vacuum indicates that the engine is under a load. The ICM will respond by retarding the timing to help prevent detonation.

Primary Ignition Components
The Ignition Control Module

The Ducellier ICM has two electrical connectors. The three wire connector carries the sig-

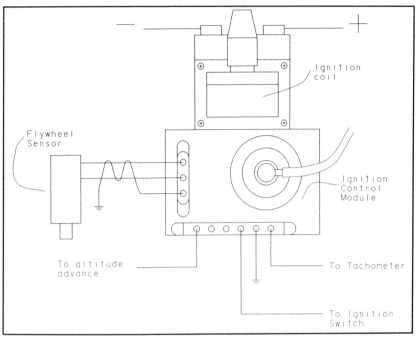

The Ducellier electronic ignition relied on computer control of ignition timing. An unusual feature for its day was the location of the pickup coil. At that time, the pickup coil was generally located in the distributor. The Ducellier system had the pickup coil referencing to the flywheel.

nals from the TDC sensor to the ICM. The second connector bank of the ICM has eight wires in three connectors that provide power and ground for the ICM as well as engine speed signals to the fuel injector computer.

Single three wire connector
- Pin A: TDC sensor
- Pin B: TDC sensor
- Pin C: not used

Eight Wire Group
Three Wire Connector #1
- Pin A: at least 9.5 volts with the engine cranking; supplies power to the ICM when the engine is cranked and when the engine is running
- Pin B: provides a ground for the ICM

The flywheel, or TDC sensor, is mounted in the bell housing on the Ducellier system. Although unique for its day in location, its operation is the same as those located in the distributor.

- Pin C: tach signal & diagnostic connector D1-Pin 1

Two Wire Connector #2
- Pin A: not used
- Pin B: tach reference signal to the ECU

Three Wire Connector #3
- No pins are used

The Ducellier ICM's capabilities are far beyond most ignition modules. In addition to switching the ignition coil on and off, it contains circuitry that uses information from the modules vacuum sensor and the TDC sensor to control the ignition timing.

Pickup coil

The pickup coil is a reluctance type pickup called a TDC sensor. Unlike most electronic ignition systems that have a distributor, the Ducellier pickup is located on the bell housing flange of the engine. Here, the pickup monitors the position and speed of the crankshaft.

The reluctor wheel of the sensor is located on the flywheel. It consists of a disc with many notches along its outer edge. Located 90deg before TDC and 90deg before Bottom Dead Center (BDC) are wider notches. The wider notches allow the ICM to keep track of the position of the crankshaft so it knows when to fire the coil.

The TDC signal feeds into the Electronic Control Unit (ECU), which is the fuel injection computer on these applications. The signal enters into the ECU on pins 11 and 28 as an AC signal. Inside the fuel injection ECU, the signal is converted into a square wave. It is used to trigger the injectors and then exits the ECU at terminal 27 to go to the ICM.

Ignition Switch

The Ducellier ignition, like most ignition systems, uses the ignition switch to control the availability of current to the ICM and ignition coil.

Secondary Ignition Components
Coil

In the Ducellier ignition system, the coil is an E core coil mounted on the ICM. The ignition coil secondary consists of hundreds of windings of very thin wire. When current stops flowing through the primary, the magnetic field created by the primary current flow collapses. This collapsing magnetic field induces several thousand volts in the secondary. This is the voltage that is used to jump the gap of the spark plugs to fire the mixture in the cylinders.

Because of the relatively low current in this high-voltage side of the coil, there are relatively few problems in the secondary side of the coil. The problems that do occur are usually in the form of open circuits.

Coil Wire

The coil secondary output wire carries the high-voltage current from the coil to the distributor cap. The coil wire is normally 6 to 12in long and has a resistance of a few thousand ohms. This relatively high resistance helps to reduce the intensity of the radio signal created by the secondary ignition.

All arcing generates a radio signal. In addition to the arcing that occurs at the spark plug, there is also an arc inside the distributor cap between the cap and the rotor. All of the secondary ignition wiring has a high resistance to reduce the affect of this arcing.

The coil wire can suffer from several possible problems. As the coil wire ages, its resistance tends to increase. Also, as the wire ages, the insulating quality of the jacket decreases, which makes it possible for the high voltages being carried by the wire to penetrate the jacket and arc to ground. Corrosion can also affect the current carrying ability of the wire.

A defective coil wire can result in misfiring, no start, and poor power.

Distributor Cap and Rotor

These two components operate as a team. The coil wire delivers the high voltage to the center terminal of the distributor cap. A carbon conductor carries the voltage to the center of the rotor. The rotor will have either a metal or carbon resistive conductor that carries the voltage to the tip of the rotor. The rotor mounts on the top of the distributor shaft and is driven by the camshaft. As the rotor rotates, it approaches either copper or aluminum conductors on the inside of the distributor cap and arcs to these conductors, which carry the voltage to the spark plug wires.

The distributor cap is prone to cracking, corrosion, and carbon tracking. Carbon tracking occurs when a microscopic crack or piece of dirt provides a current path to ground that is easier than the current path and plug wire. The rotor is subject to corrosion and perforation. Perforation occurs when the high voltage seeks and finds a ground through the rotor to the distributor shaft.

Routine replacement of the distributor cap and rotor can prevent unforeseen problems. It is not necessary to replace the cap and rotor at each tune-up as many professional technicians recommend, but they should be replaced at every other tune-up. As you read this, however, do not assume that the preceding statement means that your mechanic has been ripping you off for the past 10 years. There is no disservice in a mechanic charging you $30 or $40 extra at each tune-up to ensure that you have less chance of developing premature problems.

As discussed in Chapter 2, when replacing the distributor cap or rotor, I advise that you replace them as a set. I further advise that they be built by the same name brand manufacturer.

I have had situations in the past where a mismatched cap and rotor caused the rotor air gap to be so large that the engine either failed to start or misfired.

Spark Plug Wire

Back in the late sixties and the early seventies, when I first got into the car repair business, we checked spark plug wires by starting the engine in a darkened shop. If there were sparks flying around under the hood, it meant that one or more of the plug wires had perforated and was arcing to ground. For several years in the mid-seventies, I worked almost exclusively on fuel injected imports. Upon returning to working on domestics in 1980 I was mystified at the poor quality of the plug wires. I remembered few problems in the seventies, now there were many problems.

The difference was not the quality of the wires, but rather the leaner air/fuel ratios demanded in the early eighties (and today). In the early seventies, we unwittingly compensated for defective spark plug wires by enriching the idle mixture screws on the carburetors. This option was not available on the sealed carburetors of the late seventies. Therefore, technicians were forced to replace defective spark plug wires rather than simply masking the problem with a richer mixture.

When replacing spark plug wires, the old adage, "You get what you pay for" is especially true. A $50 set of spark plug wires can easily outlast four $12 sets of wires. A marginal plug wire can cause a misfire when the engine is under an extreme load and on a modern fuel injected car can make the engine run rich.

Spark Plugs

Every tune-up includes replacing the spark plugs. There are many brands of spark plugs on the market, some good, some bad. Asking for opinions on which is the best brand of spark plug is like asking a group which is the best soft drink, it is largely a matter of personal opinion. What I have always done, when I had a choice, was to use the brand that the manufacturer installed at the factory. My thinking is that the manufacturer has a vested interest in choosing the spark plug that would provide the best driveability and has the least chance of requiring replacement within 12,000 miles. This method has rarely failed to provide either myself or my customers with good service.

The spark plug consists of a pair of electrodes separated by an air gap of between 0.028 and 0.075in. As the spark from the coil travels down the plug wire seeking ground, it must arc across this air gap. If this arc is exposed to a properly preheated, well atomized mixture of fuel and air, this spark will ignite the fuel.

As the spark plugs arc at a high frequency in high temperatures, the electrode material will slowly vaporize. This causes the gap to widen. The wider the gap, the higher the voltage required to initiate the spark across the gap. Eventually, the voltage required to initiate the spark across the gap will be greater than the coil is capable of generating and a misfire will occur.

Common problems associated with the spark plugs are misfiring and difficulty in starting.

Timing Control

Timing control in the Ducellier ignition system is accomplished within the ICM. The pulse from the TDC sensor provides an rpm signal to the ICM. This rpm signal can be compared to the centrifugal advance of the standard ignition system. As the frequency of the TDC sensor increases, the ICM advances the timing. This parallels the effect of the weights in the centrifugal advance mechanisms of the standard point/condenser or electronic ignition system.

Changes in the ignition timing in response to changing engine load is accomplished through a vacuum sensor mounted on the ICM. During acceleration, the engine load increases and the vacuum drops. The ICM responds by retarding the ignition timing. Retarding the timing during high load conditions on the engine reduces the combustion temperatures and decreases the likelihood of detonation.

Setting the Timing

Since the TDC sensor is mounted on the engine, and since it picks up its signal from the crankshaft, the timing is not adjustable.

Troubleshooting
No Start: Basic Tests

Like most ignition systems, a no-start condition in the Ducellier ignition system is among the easiest problems to troubleshoot. There are three things required to get the engine started: air, fuel, and spark. Since the topic of this book is ignition systems, I will concentrate on the ignition problems that can prevent the engine from starting.

Begin testing the ignition system by disconnecting a plug wire from one of the spark plugs. Stick a plastic handled screwdriver in the end of one of the plug wires. While holding the screwdriver near ground, have someone crank the engine. If there is no spark, remove the coil wire from the distributor cap and place the screwdriver in the end of the coil wire. Hold the screwdriver a quarter inch from ground. Again have someone crank the engine. If there is a spark, replace the distributor rotor and cap.

If there is no spark, connect a test light to the negative terminal of the coil. While you watch the test light, have someone crank the engine. If the test light blinks, the secondary windings of the ignition coil are defective and the coil should be replaced. To verify that the secondary windings are defective, use an ohmmeter to check the resistance of the secondary. The resistance at 75deg F should be between 2,500 and 5,500 ohms. If the test light does not blink, there is a problem in the primary side of the ignition system.

Checking the TDC sensor

Begin troubleshooting the primary side of the ignition system by checking the TDC sensor. There are several viable ways to test the TDC sensor, some are more thorough than others. One of the most commonly suggested methods of testing the TDC sensor is to use an ohmmeter. The resistance between the orange and violet wire through the TDC sensor should be between 100 and 200 ohms. If the resistance through the TDC sensor is considerably different from these readings, replace it. If the resistance is less than 25 ohms outside of these specifications, the no-start problem is likely to have another cause.

The ohmmeter method of testing the TDC sensor is incomplete. While it does test the condition of the coil of wire, which is the most likely part of the TDC sensor to be defective, it does not test the condition of the permanent magnet or the air gap between the TDC sensor and the reluctor wheel. A better test is to disconnect the connector near the distributor or ignition module that contains the orange and violet wires. Connect an AC voltmeter between these wires on the distributor side of the connector. Have someone crank the engine while you watch the voltmeter. If the TDC sensor generates less than 0.5 volt, replace it. In other ignition systems, it would be important to check the air gap between the TDC sensor and the reluctor wheel. However, in the Ducellier ignition this air gap is nonadjustable. If the air gap is incorrect, it would require replacing the TDC sensor.

Testing the Module

There is no convenient way to test the ignition module directly, therefore, it is best tested through process of elimination. If a test light connected to the negative side of the coil does not blink when the engine is cranked, and if the TDC sensor tests good, replace the module.

Starts But Does Not to Run When Key Is Released

Since the Ducellier ignition system does not use a ballast resistor, the most likely cause of this symptom is a defective ignition switch.

Dies While Driving

Most commonly, this problem is caused by the ICM. However, it can also be caused by a defective TDC sensor. Test the TDC sensor by checking its resistance while heating it, cooling it, and gently tapping on it. If the reading on the meter fluctuates during the test, replace the sensor. If there is no fluctuation in the reading, replace the ICM.

Keep in mind there are many things not related to the ignition system that can cause the engine to die while the car is being driven. Among these things are the fuel injection computer (ECU) or a restricted fuel filter.

Misfire at Idle

This is almost always never the fault of the SSI system; it is almost always the fault of some component in the secondary side of the ignition system.

Before troubleshooting any misfire, it is essential to verify that the engine is in good condition. A compression test is a good starting point. If the valves are adjustable, be sure that they are properly adjusted.

With a pair of "sissy" pliers, remove and replace one plug wire at a time from the spark plugs. As each plug wire is removed, the engine rpm should drop. If one of the cylinders fails to produce as great a drop in rpm as the others, that cylinder is the source of the misfire.

Assuming the cylinder is in good condition and the valves are properly adjusted, remove the spark plug wire for that cylinder and check the resistance. The resistance should be less than 10,000 ohms per volt. If the resistance is correct, replace the spark plug. Unless the spark plugs are very new, replace them all.

Misfire Under a Load

Assuming the engine is in good condition, begin troubleshooting this problem by checking the spark plug gap. If they are gapped properly, replace the spark plugs. Even new spark plugs can misfire under a load.

If replacing the spark plugs does not solve the problem, remove the distributor cap. Inspect the wiring to the points. Frayed wiring can cause an intermittent open circuit as the vacuum advance moves the breaker plate. The intermittent open can cause a misfire.

Lack of Power

There are many things that can cause a lack of power, some related to the ignition system, some not. Begin checking this problem by confirming the engine is in good condition as are the air and fuel filters.

If a lack of power is the result of problems in the ignition system, it is likely the problem is in the timing control system. To test the timing control system, connect a timing light to the engine. Disconnect the vacuum advance and plug in the hose. With the engine at idle speed, check the timing. Now raise the engine speed to 2000 to 2500 rpm. If the timing does not advance, replace the ICM.

Chrysler Passenger Cars and Light-Duty Trucks

7

Chrysler introduced its first electronic ignition system into mass production in 1972, although it was offered as an option in 1971. The introduction of this system coincided with my entry into the auto repair business. I remember at the time hearing my colleagues comment that the new technology was just a passing fad. I remember them saying that it was an impractical system. I guess time has proved them wrong. From 1972 to 1987, three variations on this system were made.

Primary Ignition Components
Control Module

The Control Module consists of a large power transistor that grounds the negative terminal of the coil. This transistor is mounted on the outside of the module to improve its cooling. In the early days, these units had a fairly high failure rate, or so it seemed. Perhaps many of those early problems were really a failure to understand the operation of the system and a failure to understand the effects of corrosion. A technician would often replace the ignition module out of desperation. When the old module was disconnected, the corrosion that had built up on the connections would be cleaned off. Installing the new module would further clean the connections so that the fresh module would start the car.

There are five wires, arranged in a star shape, connected to the ignition module on the 1972-1979 Chrysler electronic ignition systems. With the module held so that the power transistor is toward the bottom, the two wires on the right (wire numbers 4 and 5) go to the pickup coil in the distributor. Each of the two wires should have virtually no resistance in them between the module connector and the connection at the distributor. Measured between these two ignition module terminals with the wires still connected to the distributor, the resistance should be between 150 and 900 ohms. When the resistance of these wires is measured in this manner, it measures not only the wires, but the connection and the pickup coil as well.

Continuing clockwise around the connector wire, number 3 (the lower left) is connected to the 5 ohm leg of the dual ballast resistor. This wire is the voltage source for the power transistor. The upper left corner of the star is terminal 2. Connected to the negative terminal on the coil, this is the wire that the power transistor uses to switch the ignition coil on and off. The top wire, wire number 1, connects to the ballast resistor and carries full battery voltage to the ignition module all the time. This is the main power supply for the ignition module. The ignition module is case ground to the chassis.

In the models produced from 1980 through 1987, the number 3 wire (lower left) is missing. The power transistor receives its power from the main ignition module power feed at terminal 1.

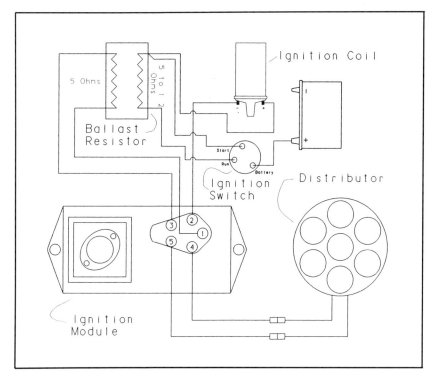

Chrysler was the first US manufacturer to introduce electronic ignition in 1971. The system proved to be dependable, but scared many professional technicians out of the business.

57

The rest of the terminals perform their function just as they did on the earlier models. From 1981 through 1987, some applications had dual pickups in the distributor. These are easily recognized as there are four wires going into the base of the distributor rather than two. Between the distributor and the ignition module there is a Dual Pickup Start Relay. The starter relay is energized by turning the ignition switch to the start position. At this point the Dual Pickup Start Relay causes the ignition module to receive engine speed signals from the run pickup coil instead of the start pickup coil. Both the dual pickup and single pickup models later than 1980 use a single element ballast resistor.

Pickup Coil

The pickup coil consists of a coil of wire positioned in a magnetic field from a permanent magnet. Opposite the coil of wire, mounted on the distributor shaft, is a wheel with a series of reluctor teeth around its circumference. As the teeth rotate, they pass through the magnetic field, warping the magnetic field across the coil of wire. This induces a voltage in the coil of wire. The approaching tooth warps the magnetic field in the direction of its approach. A voltage is induced. As the distributor shaft continues to rotate, it drags the magnetic field with it. Eventually, the tooth passes its point of closest approach with the pickup and begins to move away. The magnetic field is dragged in the opposite direction across the coil of wire. A voltage is then produced in the opposite direction. The resulting output of the pickup is an alternating current, the voltage output and frequency of which are directly proportional to the rotational speed of the distributor shaft.

Ignition Switch

The ignition switch controls whether current will be available to the coil and ignition module. Like the point/condenser ignition system, the Chrysler electronic system uses a different current path when starting the engine than it uses when the engine is running.

Secondary Ignition Components
Coil

The ignition coil secondary consists of hundreds of windings of very thin wire. When current stops flowing through the primary, the magnetic field created by the primary current flow collapses. This collapsing magnetic field induces several thousand volts in the secondary. This is the voltage that is used to jump the gap of the spark plugs to fire the mixture in the cylinders.

Because of the relatively low current in this high-voltage side of the coil, there are relatively few problems in the secondary side of the coil. The problems that do occur are usually in the form of open circuits.

Coil Wire

The coil secondary output wire carries the high-voltage current from the coil to the distributor cap. The coil wire is normally 6 to 12in long and has a resistance of a few thousand ohms. This relatively high resistance helps to reduce the intensity of the radio signal created by the secondary ignition.

All arcing generates a radio

The Chrysler ignition system of the early seventies, as well as many later applications, used the standard pickup coil. The pickup coil air gap is adjustable.

The standard, oil-filled ignition coil remained in use by Chrysler until the nineties.

signal. In addition to the arcing that occurs at the spark plug, there is also an arc inside the distributor cap between the cap and the rotor. All of the secondary ignition wiring has a high resistance to reduce the affect of this arcing.

The coil wire can suffer from several possible problems. As the coil wire ages its resistance tends to increase. Also, as the wire ages the insulating quality of the jacket decreases, which makes it possible for the high voltages being carried by the wire to penetrate the jacket and arc to ground. Corrosion can also affect the current carrying ability of the wire.

A defective coil wire can result in misfiring, no-start, and poor power.

Distributor Cap and Rotor

These two components operate as a team. The coil wire delivers the high voltage to the center terminal of the distributor cap. A carbon conductor carries the voltage to the center of the rotor. The rotor will have either a metal or carbon resistive conductor that carries the voltage to the tip of the rotor. The rotor mounts on the top of the distributor shaft and is driven by the camshaft. As the rotor rotates, it approaches either copper or aluminum conductors on the inside of the distributor cap and arcs to these conductors, which carry the voltage to the spark plug wires.

The distributor cap is prone to cracking, corrosion, and carbon tracking. Carbon tracking occurs when a microscopic crack or piece of dirt provides a current path to ground that is easier than the current path and plug wire. The rotor is subject to corrosion and perforation. Perforation occurs when the high voltage seeks and finds a ground through the rotor to the distributor shaft.

Routine replacement of the distributor cap and rotor can prevent unforeseen problems. It is not necessary to replace the cap and rotor at each tune-up as many professional technicians recommend, but they should be replaced at every other tune-up. As you read this, however, do not assume that the preceding statement means that your mechanic has been ripping you off for the past 10 years. There is no disservice in a mechanic charging you $30 or $40 extra at each tune-up to ensure that you have less chance of developing premature problems.

As discussed in Chapter 2, when replacing the distributor cap or rotor, I advise that you replace them as a set. I further advise that they be built by the same name brand manufacturer. I have had situations in the past where a mismatched cap and rotor caused the rotor air gap to be so large that the engine either failed to start or misfired.

Spark Plug Wires

Back in the late sixties and the early seventies, when I first got into the car repair business, we checked spark plug wires by starting the engine in a darkened shop. If there were sparks flying around under the hood, it meant that one or more of the plug wires had perforated and was arcing to ground. For several years in the mid-seventies, I worked almost exclusively on fuel injected imports. Upon returning to working on domestics in 1980, I was mystified at the poor quality of the plug wires. I remembered few problems in the seventies, now there were many problems.

The difference was not the quality of the wires, but rather the leaner air/fuel ratios demanded in the early eighties (and today). In the early seventies, we unwittingly compensated for defective spark plug wires by enriching the idle mixture screws on the carburetors. This option was not available on the sealed carburetors of the late seventies. Therefore, technicians were forced to replace defective spark plug wires rather than simply masking the problem with a richer mixture.

When replacing spark plug wires, the old adage, "You get what you pay for" is especially true. A $50 set of spark plug wires can easily outlast four $12 sets of wires. A marginal plug wire can cause a misfire when the engine is under an extreme load and on a modern fuel injected car can make the engine run rich.

Spark Plugs

Every tune-up includes replacing the spark plugs. There are many brands of spark plugs on the market, some good, some bad. Asking for opinions on which is the best brand of spark plug is like asking a group which is the best soft drink, it is largely a matter of personal opinion. What I have always done, when I had a choice, was to use the brand that the manufacturer installed at the factory. My thinking is that the manufacturer has a vested interest in choosing the spark plug that would provide the best driveability and has the least chance of requiring replacement within 12,000 miles. This method has rarely failed to provide either myself or my customers with good service.

The spark plug consists of a pair of electrodes separated by an air gap of between 0.028 and 0.075in. As the spark from the coil travels down the plug wire seeking ground, it must arc across this air gap. If this arc is exposed to a properly preheated, well atomized mixture of fuel and air, this spark will ignite the fuel.

As the spark plugs arc at a high frequency in high temperatures, the electrode material will slowly vaporize. This causes the gap the widen. The wider the gap, the higher the voltage required to initiate the spark across the gap. Eventually, the

59

voltage required to initiate the spark across the gap will be greater than the coil is capable of generating and a misfire will occur.

Common problems associated with the spark plugs are misfiring and difficulty in starting.

Timing Control
Centrifugal Advance

As an example, on a hypothetical engine the spark occurs at 10deg before TDC when the engine is running at 1000rpm. At this speed, the spark extinguishes at about 10deg after TDC. This means that the crankshaft has rotated 20deg since the initiation of spark. As the engine speed increases, the crankshaft rotates more degrees in the 2.5 milliseconds that the spark is jumping the gap. At 2000rpm the crankshaft will rotate twice as much. If the timing at 1000rpm should be 10deg before TDC, then the timing at 2000rpm should be about 30deg before TDC. As the engine speed continues to increase, the timing will need to continue to advance. The amount of total advance, the upper limit of the advance, will vary depending on the design of the engine.

The change of timing in response to rpm is accomplished through a set of spring-loaded weights. As the speed of the engine increases, the weights swing out against spring tension. The trigger wheel that trips the sensor, although it is mounted on the distributor shaft, is not part of the distributor shaft. The swinging weights cause the trigger wheel to rotate with respect to the distributor shaft. This advances the timing.

Vacuum Advance

At first glance this is a misnomer. The vacuum advance actually retards the timing when the engine is under a load. In most applications the vacuum advance is connected to ported vacuum. The advance unit receives no vacuum at an idle, but when the throttle is opened the vacuum advances the timing. As the load on the engine increases, the vacuum drops. As the vacuum drops, the timing is not advanced as much, it retards. Retarding the timing lowers the combustion temperature, and therefore prevents detonation and decreases the potential of damage to the engine.

Setting the Timing

When setting the ignition timing, it is best to follow the instructions outlined on the EPA sticker under the hood. If the instructions are missing or obliterated, disconnect the vacuum advance and adjust to the specification listed in this book.

Troubleshooting
No Start: Basic Tests

Electronic ignition systems require the same basic tests as the old point/condenser system. Like most ignition systems, a no-start condition in 1972-1987 Chrysler electronic ignition systems is among the easiest problems to troubleshoot. There are three things required to get the engine started: air, fuel, and spark. Since the topic of this book is ignition systems, I will concentrate on the ignition problems that can prevent the engine from starting.

Begin testing the ignition system by disconnecting a plug wire from one of the spark plugs. Stick a plastic handled screwdriver in the end of one of the plug wires. While holding the screwdriver near ground, have someone crank the engine. If there is no spark, remove the coil wire from the distributor cap and place the screwdriver in the end of the coil wire. Hold the screwdriver about a quarter inch from ground. Again, have someone crank the engine. If there is a spark, replace the distributor rotor and cap.

Using an engine ignition analyzer, such as the one shown here, helps in troubleshooting ignition problems. Time and careful replacement of components can usually accomplish the same thing. The analyzer's only real advantage is that it makes the troubleshooting procedure faster and easier for the professional technician.

If there is no spark, connect a test light to the negative terminal of the coil. While you watch the test light, have someone crank the engine. If the test light blinks, the secondary windings of the ignition coil are defective and the coil should be replaced. To verify that the secondary windings are defective, use an ohmmeter to check the resistance of the secondary. The resistance should be between 8,000 and 12,000 ohms. If the test light does not blink, there is a problem in the primary side of the ignition system.

Checking the pickup coil

Begin troubleshooting the primary side of the ignition system by checking the pickup coil. There are several viable ways to test the pickup coil, some are more thorough than others. One of the most commonly suggested methods of testing the pickup coil is to use an ohmmeter. The resistance between the two wires (or each pair if the vehicle is later than a 1981 and has dual pickups) running between terminals 4 and 3 of the pickup coil should be between 150 and 900 ohms. If the resistance through the pickup coil is considerably different from these readings, replace it. If the resistance is less than 100 ohms outside of these specifications, the no-start problem is likely to have another cause.

The ohmmeter method of testing the pickup coil is incomplete. While it does test the condition of the coil of wire, which is the most likely part of the pickup coil to be defective, it does not test the condition of the permanent magnet or the air gap between the pickup and the reluctor wheel. A better test is to disconnect the connector near the distributor or ignition module that contains the two wires that run between the ignition module and the distributor. Connect an AC voltmeter between these wires on the distributor side of

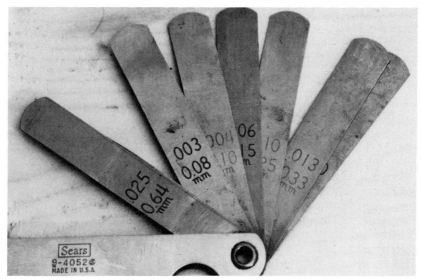

Pickup coil air gap is as critical on the Chrysler electronic ignition system as it is in a point/condenser ignition system. For the sake of accuracy, use a non-ferrous feeler gauge when checking this gap. Steel feeler gauges will be attracted by the magnet in the pickup coil and stick to the pickup coil, and this makes accurate adjustment difficult.

the connector. Have someone crank the engine while you watch the voltmeter. If the pickup generates less than 0.5 volt, replace the pickup. Check the air gap between the pickup and the reluctor wheel and adjust as necessary.

Testing the Module

There is no convenient way to test the ignition module directly, therefore it is best tested through process of elimination. If there is power to the positive side of the coil and a test light connected to the negative side of the coil does not blink when the engine is cranked, and if the pickup tested good, replace the module.

Starts But Does Not Continue to Run When the Key Is Released

In each of the three variations on the Chrysler electronic, a ballast resistor is used to limit the flow of current through the ignition primary. While there are other possible causes of this problem, a defective ballast resistor ranks as one of the most likely causes. When the key is in the crank position, current flows to the ignition coil directly from the battery to the ignition switch. This allows full battery voltage to the coil. As important as the voltage is, that full current is applied to the coil while cranking is limited only by the coil itself. Full battery voltage and full current allow the maximum power from the coil while the engine is being started. When the starter is engaged, the battery voltage is reduced due to the load of the starter; this reduces the watts of energy available to the ignition system. If the ballast resistor were left in the circuit during start-up, the wattage could be too low to create a sufficient spark to start the engine. When the engine is running, the current to the coil passes through the ballast resistor. This limits the current and prevents the primary circuit from overheating and limits the secondary output voltage and current. If the ballast resistor is burned out, the current can get to the coil when the engine is being cranked but not when the key is released to the run position.

Dies While Driving Down the Road

The most likely cause of this condition is loose or corroded connections. Inspect the primary ignition connections thoroughly all the way from the battery to the distributor. This is, of course, easier said than done since many of the wires and connections are under the dash. A little personal judgment may be appropriate here. Is it worth taking a chance buying parts you may not need, or do you prefer the piece of mind in knowing that you really needed the part you bought before you buy it? Keep in mind that the ignition switch is a possible cause of this problem. If the switch feels loose or if wiggling the key gently in the switch causes the engine to die, replace the ignition switch.

Testing the Distributor Pickup

A likely cause of intermittent stalling while the vehicle is moving down the road is the pickup coil. Vibration or heat can cause intermittent open circuits to occur in the windings of the pickup. To test the pickup for this condition, remove the distributor cap and connect an ohmmeter across the leads of the pickup as they go into the distributor. Observe the reading on the ohmmeter. The resistance should be between 150 and 900 ohms. If the reading is outside of this specification, replace the pickup. If the reading is acceptable, heat the pickup with a light bulb for several minutes. The reading should only have changed very slightly. If the reading did not change, tap on the pickup with the handle end of a screwdriver. If the reading fluctuates, replace the pickup. If they do not fluctuate, replace the ignition module.

Testing the Module

While there is no accurate test for the condition of the ignition module, process of elimination can be effective. If the connections and wires are all in good condition, and if the tests on the pickup coil reveal no problems, replace the ignition module. Of course, the surefire cure, once the ignition switch and the wires have been confirmed, is to replace both the ignition module and the pickup coil.

Misfire at Idle

This is almost always never the fault of the Chrysler electronic ignition system; it is almost always the fault of some component in the secondary side of the ignition system.

Before troubleshooting any misfire, it is essential to verify that the engine is in good condition. A compression test is a good starting point. If the valves are adjustable, be sure that they are properly adjusted.

With a pair of "sissy" pliers, remove and replace one plug wire at a time from the spark plugs. As each plug wire is removed, the engine rpm should drop. If one of the cylinders fails to produce as great a drop in rpm as the others, that cylinder is the source of the misfire.

Assuming the cylinder is in good condition and the valves are properly adjusted, remove the spark plug wire for that cylinder and check the resistance. The resistance should be less than 10,000 ohms per volt. If the resistance is correct, replace the spark plug. Unless the spark plugs are very new, replace them all.

Misfire Under a Load

Assuming the engine is in good condition, begin troubleshooting this problem by checking the spark plug gap. If they are gapped properly, replace the spark plugs. Even new spark plugs can misfire under a load.

If replacing the spark plugs does not solve the problem, remove the distributor cap. Inspect the wiring to the points. Frayed wiring can cause an intermittent open circuit as the vacuum advance moves the breaker plate. The intermittent open can cause a misfire.

Lack of Power

There are many things that can cause a lack of power, some related to the ignition system, some not. Begin checking this problem by confirming the engine is in good condition as are the air and fuel filters.

If a lack of power is the result of problems in the ignition system, it is likely the problem is in the timing control system. To test the timing control system, connect a timing light to the engine. Disconnect the vacuum advance and plug in the hose. With the engine at idle speed, check the timing. Now raise the engine speed to 2000 to 2500rpm. If the timing does not advance, the centrifugal advance system is not working. Inspect the distributor weights. If they are free and move easily, replace the weight springs. If the springs are weak, they will allow the timing to advance all the way prematurely, even at idle. If the weights are frozen, use penetrating oil or whatever is necessary to free them. If they are badly corroded, it may be necessary to replace the distributor.

If, or when, the centrifugal advance is working properly, with the engine still at 2000 to 2500rpm, reconnect the vacuum hose to the vacuum advance. When the vacuum hose is reconnected, the timing should advance several degrees.

Chrysler Hall Effects Ignition

Some 1980 Omnis and Horizons, those meeting federal emission standards, used an ignition system similar to that used on other models that year. The most significant difference is the use of a Hall Effects pickup rather than a pickup coil.

Primary Ignition Components
Electronic Control Unit

The ECU features a large externally mounted power transistor with five wires connected to the ignition module. A dark blue wire is connected to the output side of the ignition switch. On the ignition module, it is connected to terminal 2. This same wire also supplies power to the positive terminal of the ignition coil. There is a black/yellow wire that runs from the negative terminal of the ignition coil to terminal 5 of the ignition module. The other three wires connect the ignition module to the Hall Effects. The wires form a triangle in the connector from the distributor to the ignition module.

Numbered clockwise, the top wire is number 1 and corresponds to terminal 1 on the ignition module. This is the power lead to the Hall Effects. The lower right, or number 2 terminal in the distributor connector, corresponds to terminal 4 of the ignition module. This is the ground for the Hall Effects. Terminal 3 of the distributor connector is also terminal 3 on the ignition module. This carries the distributor reference signal from the Hall Effects to the module.

Pickup Coil (Hall Effects)

The primary advantage of a Hall Effects sensor over the pickup coil is its ability to detect position and rotational speed from zero rpm to tens of thousands. Its primary disadvantage is that it is not as rugged as the pickup coil and is more sensitive to errant magnetic fields. An intense magnetic field can shut down the proper operation of a Hall Effects.

How the Hall Effects Works

A Hall Effects pickup is a semiconductor carrying a current. When a magnetic field falls perpendicular to the direction of that current flow, part of that current is redirected perpendicular to the main current path. The semiconductor is placed near a permanent magnet. A set of metal blades, or armature, attached to a rotating shaft or other device passes between the Hall Effects semiconductor and the permanent magnet. As the armature rotates, the magnet field is alternately applied to the Hall Effects and interrupted. The result is a pulsing current perpendicular to the main current path. This frequency is directly proportional to the speed of armature rotation. Since the output is only dependent on the presence of the magnetic field, the Hall Effects unit is capable of detecting armature position even when there is no rotational speed.

Ignition Switch

The ignition switch controls whether current will be available to the coil and ignition module. Unlike the point/condenser

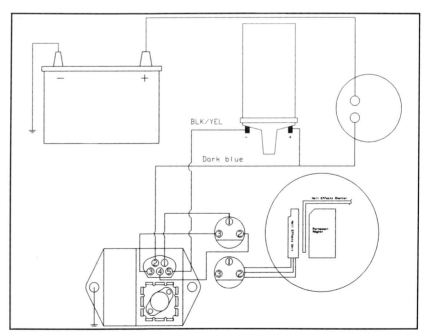

In 1980, Chrysler used a Hall Effects ignition system on the federal Omnis and Horizons. Essentially, the system bore many resemblances to the regular Chrysler ignition system. The only significant difference is the use of a Hall Effects rather than a pickup coil.

and standard Chrysler electronic ignition systems, the Chrysler Hall Effects system uses the same current path when starting the engine that it uses when the engine is running. There is no ballast resistor in this system.

Secondary Ignition Components
Coil

The ignition coil secondary consists of hundreds of windings of very thin wire. When current stops flowing through the primary, the magnetic field created by the primary current flow collapses. This collapsing magnetic field induces several thousand volts in the secondary. This is the voltage that is used to jump the gap of the spark plugs to fire the mixture in the cylinders.

Because of the relatively low current in this high-voltage side of the coil, there are relatively few problems in the secondary side of the coil. The problems that do occur are usually in the form of open circuits.

Coil Wire

The coil secondary output wire carries the high-voltage current from the coil to the distributor cap. The coil wire is normally 6 to 12in long and has a resistance of a few thousand ohms. This relatively high resistance helps to reduce the intensity of the radio signal created by the secondary ignition.

All arcing generates a radio signal. In addition to the arcing that occurs at the spark plug, there is also an arc inside the distributor cap between the cap and the rotor. All of the secondary ignition wiring has a high resistance to reduce the affect of this arcing.

The coil wire can suffer from several possible problems. As the coil wire ages its resistance tends to increase. Also, as the wire ages the insulating quality of the jacket decreases, which makes it possible for the high voltages being carried by the wire to penetrate the jacket and arc to ground. Corrosion can also affect the current carrying ability of the wire.

A defective coil wire can result in misfiring, no-start, and poor power.

Distributor Cap and Rotor

These two components operate as a team. The coil wire delivers the high voltage to the center terminal of the distributor cap. A carbon conductor carries the voltage to the center of the

The Hall Effects sensor used in the Chrysler Hall Effects ignition system creates a square wave as the shutter blades pass through the gap between the semiconductor and the permanent magnet. The sensor can be tested with a voltmeter, tachometer, or dwell meter.

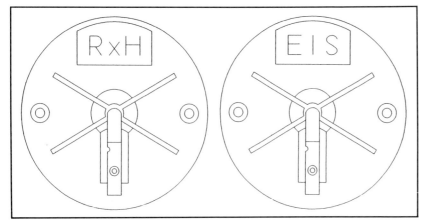

When replacing the rotor in the Chrysler Hall Effects ignition system, be sure to use the correct one. Although these rotors look basically alike, they are not interchangeable.

rotor. The rotor will have either a metal or carbon resistive conductor that carries the voltage to the tip of the rotor. The rotor mounts on the top of the distributor shaft and is driven by the camshaft. As the rotor rotates, it approaches either copper or aluminum conductors on the inside of the distributor cap and arcs to these conductors, which carry the voltage to the spark plug wires.

The distributor cap is prone to cracking, corrosion, and carbon tracking. Carbon tracking occurs when a microscopic crack or piece of dirt provides a current path to ground that is easier than the current path and plug wire. The rotor is subject to corrosion and perforation. Perforation occurs when the high voltage seeks and finds a ground through the rotor to the distributor shaft.

Routine replacement of the distributor cap and rotor can prevent unforeseen problems. It is not necessary to replace the cap and rotor at each tune-up as many professional technicians recommend, but they should be replaced at every other tune-up. As you read this, however, do not assume that the preceding statement means that your mechanic has been ripping you off for the past 10 years. There is no disservice in a mechanic charging you $30 or $40 extra at each tune-up to ensure that you have less chance of developing premature problems.

As discussed in Chapter 2, when replacing the distributor cap or rotor, I advise that you replace them as a set. I further advise that they be built by the same name brand manufacturer. I have had situations in the past where a mismatched cap and rotor caused the rotor air gap to be so large that the engine either failed to start or misfired.

Spark Plug Wires

Back in the late sixties and the early seventies, when I first got into the car repair business, we checked spark plug wires by starting the engine in a darkened shop. If there were sparks flying around under the hood, it meant that one or more of the plug wires had perforated and was arcing to ground. For several years in the mid-seventies, I worked almost exclusively on fuel injected imports. Upon returning to working on domestics in 1980, I was mystified at the poor quality of the plug wires. I remembered few problems in the seventies, now there were many problems.

The difference was not the quality of the wires, but rather the leaner air/fuel ratios demanded in the early eighties (and today). In the early seventies, we unwittingly compensated for defective spark plug wires by enriching the idle mixture screws on the carburetors. This option was not available on the sealed carburetors of the late seventies. Therefore, technicians were forced to replace defective spark plug wires rather than simply masking the problem with a richer mixture.

Spark Plugs

Every tune-up includes replacing the spark plugs. There are many brands of spark plugs on the market, some good, some bad. Asking for opinions on which is the best brand of spark plug is like asking a group which is the best soft drink, it is largely a matter of personal opinion. What I have always done, when I had a choice, was to use the brand that the manufacturer installed at the factory. My thinking is that the manufacturer has a vested interest in choosing the spark plug that would provide the best driveability and has the least chance of requiring replacement within 12,000 miles. This method has rarely failed to provide either myself or my customers with good service.

The spark plug consists of a pair of electrodes separated by an air gap of between 0.028 and 0.075in. As the spark from the coil travels down the plug wire seeking ground, it must arc across this air gap. If this arc is exposed to a properly preheated, well atomized mixture of fuel and air, this spark will ignite the fuel.

As the spark plugs arc at a high frequency in high temperatures, the electrode material will slowly vaporize. This causes the gap to widen. The wider the gap, the higher the voltage required to initiate the spark across the gap. Eventually, the voltage required to initiate the spark across the gap will be greater than the coil is capable of generating and a misfire will occur.

Common problems associated with the spark plugs are misfiring and difficulty in starting.

Timing Control
Centrifugal Advance

A standard set of weights is used to control the ignition timing in response to changes in rpm. The design differs little from the design in a standard point/condenser ignition system. Defective or frozen weights can result in low power.

Vacuum Advance

Timing control in response to changes in engine load is accomplished through a standard vacuum advance unit. When the load on the engine is low, the vacuum is high; the vacuum advance pulls the Hall Effects to the fully advanced position. When the engine comes under a load, the vacuum drops and the timing retards. If the timing fails to retard properly when the engine load increases, the engine will likely detonate.

Setting the Timing

When setting the ignition timing, it is best to follow the instructions outlined on the EPA sticker under the hood. If the instructions are missing or obliterated, disconnect the vacuum advance and adjust to the specification listed in this book.

Troubleshooting
No Start: Basic Tests

Testing the distributor pickup; testing the Hall Effects:

With an Ohmmeter

There is no valid test procedure on the Hall Effects using an ohmmeter.

With an Oscilloscope

Connect the oscilloscope to the Hall Effects signal lead. Rotate the armature. Depending on the number of blades and the rotational speed of the armature, the scope pattern could appear either as a square wave or a flat line that rises and falls with rotation.

With a Voltmeter

Connect a voltmeter to the Hall Effects output lead. The voltmeter should display either a digital high (4 volts or more) or a digital low (around 0 volt). Slowly rotate the armature while observing the voltmeter. If the voltmeter had read low, it should now read high; if the voltmeter had read high, it should now read low. If the voltage fluctuates in this manner as the armature is rotated, then the Hall Effects is good.

With a Dwell Meter

Since the signal generated by the Hall Effects is a square wave, the dwell meter becomes a natural for testing. Connect the dwell meter between the Hall Effects output and ground. Rotate the armature as fast as possible (example, crank the engine), he dwell meter should read something besides zero and full scale. If it does, the Hall Effects is good.

With a Tachometer

As with the dwell meter, the tachometer is also a good tool for detecting square wave. Connect the tachometer between the Hall Effects output and ground. With the armature rotating as described in the paragraph on the dwell meter, the tachometer should read something other than zero if the Hall Effects is good.

Testing the Module

Many technicians are frustrated by the fact that they cannot accurately test the condition of the ignition module. The author sympathizes with this frustration. Replacing a $100 component that is not confirmed to be defective is an agonizing experience. The agony is compounded by the fact that most auto parts stores will not accept returns on ignition modules if it turns out that the defective ignition module was a misdiagnosis. Unfortunately, especially in light of all of the aforementioned, there is no way to accurately diagnose an ignition module. The only way is by process of elimination; the problem is that there always is a possibility of overlooking one of the potential causes.

Starts But Does Not Continue to Run When the Key Is Released

Since this system uses no ballast resistor, the most likely cause of this symptom is a defective ignition switch. Be sure the spark plugs are in good condition before replacing the switch. Occasionally, a set of spark plugs will allow the engine to start but will not supply an intense enough spark to keep the engine running. When the engine is started, the typical owner of a carbureted car will pump the accelerator several times prior to turning the key. This provides the combustion chambers with an extremely rich mixture to start the engine. In this rich environment, the spark plugs have an easy task of igniting the mixture. As soon as the engine starts, the mixture leans out and the engine dies.

If the spark plugs are new, or in good condition, connect a test light to pin 2 of the ignition module and start the engine. If the test light goes out as the engine dies, inspect the wires and connections from the ignition switch to the coil and ignition module carefully. If the wires

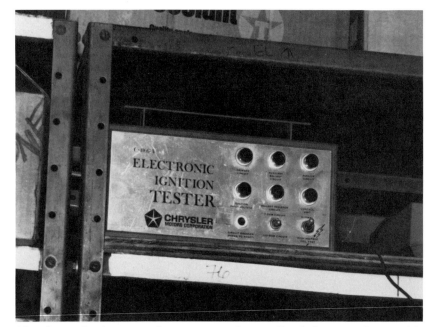

Chrysler and other manufacturers built and sold special testers for their electronic ignition systems. Most of them suffered the same fate as this unit, sentenced to spend eternity on the shelf in the back of the parts room.

and connections are in good condition, replace the ignition switch. If the test light remains on while the engine dies, replace the ignition module.

Dies While Driving Down the Road

Testing the distributor pickup: The Hall Effects pickup is notorious for intermittent failure—which is always frustrating. One of the imps of ancient Greek mythology is assigned to the creation of intermittent car problems. As a sentient being, this imp knows when a technician or car owner is looking for the source of the intermittent fault and allows the car to perform perfectly. As soon as the car gets into heavy, freeway traffic, the imp allows the fault to return.

While a Hall Effects can be easily tested for a no-start condition, testing it for an intermittent failure is virtually impossible. There are no valid resistance checks, the only way to test a suspected Hall Effects is to replace it and see if the symptom disappears. Yes, as you have already recognized, this is the classic "replace with known good unit" diagnostic technique. To add to the reader's frustration, this is exactly the same procedure that must be used to troubleshoot the ignition module for intermittents.

Before replacing either the ignition module or the Hall Effects, it must be confirmed that several other components are in good condition. Confirm the ignition switch is in good condition using the technique discussed above. Check the wires and connections to the ignition module, distributor, and coil. If they are in good condition, connect a test light to the number 2 terminal of the ignition module. Position the test light so you can see it as you drive down the road. When the car dies, is the test light on? If so, replace either the Hall Effects or the ignition module. If the light goes out, repair the intermittent open, short, or ground.

If the wiring test results indicate the problem is not in the wiring, you must make a decision. The problem is either in the ignition module, the Hall Effects, or in the wiring between them. Inspect the wiring carefully. If the wiring is in good condition, call a few parts houses. The price of the ignition module and the Hall Effects is usually about the same, within ten dollars of one another. Since the ignition module is much easier to replace, change it first. If the problem persists, replace the Hall Effects.

Misfire at Idle

This is almost always never the fault of the Hall Effects ignition system; it is almost always the fault of some component in the secondary side of the ignition system.

Before troubleshooting any misfire, it is essential to verify that the engine is in good condition. A compression test is a good starting point. If the valves are adjustable, be sure that they are properly adjusted.

With a pair of "sissy" pliers, remove and replace one plug wire at a time from the spark plugs. As each plug wire is removed, the engine rpm should drop. If one of the cylinders fails to produce as great a drop in rpm as the others, that cylinder is the source of the misfire.

Assuming the cylinder is in good condition and the valves are properly adjusted, remove the spark plug wire for that cylinder and check the resistance. The resistance should be less than 10,000 ohms per volt. If the resistance is correct, replace the spark plug. Unless the spark plugs are very new, replace them all.

Misfire Under a Load

Assuming the engine is in good condition, begin troubleshooting this problem by checking the spark plug gap. If they are gapped properly, replace the spark plugs. Even new spark plugs can misfire under a load.

If replacing the spark plugs does not solve the problem, remove the distributor cap. Inspect the wiring to the points. Frayed wiring can cause an intermittent open circuit as the vacuum advance moves the breaker plate. The intermittent open can cause a misfire.

Lack of Power

There are many things that can cause a lack of power, some related to the ignition system, some not. Begin checking this problem by confirming the engine is in good condition as are the air and fuel filters.

If a lack of power is the result of problems in the ignition system, it is likely the problem is in the timing control system. To test the timing control system, connect a timing light to the engine. Disconnect the vacuum advance and plug in the hose. With the engine at idle speed, check the timing. Now raise the engine speed to 2000 to 2500rpm. If the timing does not advance, the centrifugal advance system is not working. Inspect the distributor weights. If they are free and move easily, replace the weight springs. If the springs are weak, they will allow the timing to advance all the way prematurely, even at idle. If the weights are frozen, use penetrating oil or whatever is necessary to free them. If they are badly corroded, it may be necessary to replace the distributor.

If, or when, the centrifugal advance is working properly, with the engine still at 2000 to 2500 RPM, reconnect the vacuum hose to the vacuum advance. When the vacuum hose is reconnected, the timing should advance several degrees.

Mitsubishi Electronic Ignition System 9

The Mitsubishi electronic ignition system was used on Chrysler Corporation Caravans, Ram Vans, and Voyagers from 1985 through 1987, as well as passenger cars from 1981 through 1985. Unique to the 2.6 liter engine, the system consists of the battery, ignition switch, ignition coil, ECU (here called an IC ignitor), electronic speed sensor, starter relay, and secondary ignition components. There is no ballast resistor in the system.

Primary Ignition Components
Speed Sensor

The speed sensor works like a governor. When the sensor detects that the pulses at the negative terminal of the coil indicate an engine speed greater than a specified amount, the speed sensor cuts off power to the positive terminal of the ignition coil. In addition, power is cut off to the carburetor bowl vent solenoid, the fuel cutoff solenoid, the air pump control solenoid, and the deceleration spark advance solenoid. Essentially, the engine is shut down until the ignitor indicates to the speed sensor an engine speed detected from the pickup coil below the specified

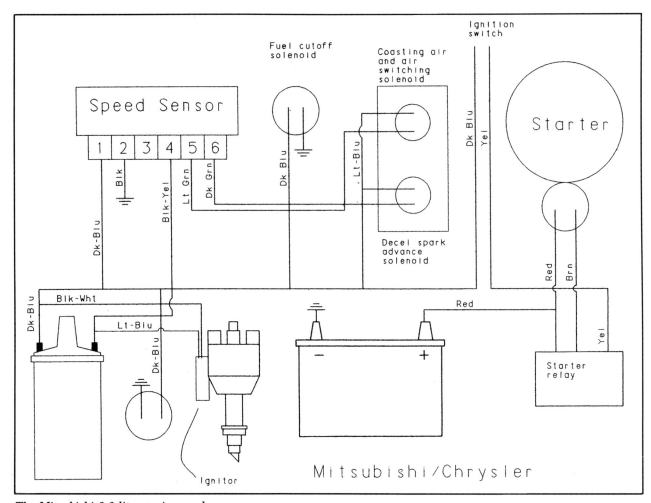

The Mitsubishi 2.6 liter engine used in the early Dodge Caravan used a Mitsubishi ignition system.

amount. This system prevents over-revving and the possibility of damaging the engine.

Electronic Control Unit

The ECU is an integrated circuit ignition module located in the distributor. Often referred to as an ignitor, the module translates the sine wave AC voltage signals of the pickup coil into square waves that turn the ignition coil on and off.

Pickup Coil

In the Mitsubishi electronic ignition system, the pickup coil is a reluctance pickup. This sensor produces an alternating current as the distributor rotates. The sensor consists of a coil of wire, a permanent magnet, and a reluctor wheel. The reluctor wheel is part of the distributor shaft. As the reluctor wheel rotates, it distorts the magnetic field back and forth across the coil of wire. This distorting induces an alternating current in the coil of wire. As the speed of distributor shaft rotation increases, the amplitude (voltage output) of the sensor increases. More important, since the AC signal is a sine wave, the frequency of the sine wave changes as the speed of the distributor shaft changes. The frequency of the sine wave tells the ignition module the speed and position of the distributor shaft. Since the distributor shaft is synchronized to the cam shaft, the ignition is synchronized to the mechanical components of the engine. The ECU uses the signal from the pickup coil as a trigger source for the firing of the coil.

Ignition Switch

The ignition switch controls whether current will be available to the coil and ignition module. There is no ballast resistor in this system. The current through the primary is limited by circuitry in the ignition module.

Secondary Ignition Components
Coil

The ignition coil secondary consists of hundreds of windings of very thin wire. When current stops flowing through the primary, the magnetic field created by the primary current flow collapses. This collapsing magnetic field induces several thousand volts in the secondary. This is the voltage that is used to jump the gap of the spark plugs to fire the mixture in the cylinders.

Because of the relatively low current in this high-voltage side of the coil, there are relatively few problems in the secondary side of the coil. The problems that do occur are usually in the form of open circuits.

Coil Wire

The coil secondary output wire carries the high-voltage current from the coil to the distributor cap. The coil wire is normally 6 to 12in long and has a resistance of a few thousand ohms. This relatively high resistance helps to reduce the intensity of the radio signal created by the secondary ignition.

All arcing generates a radio signal. In addition to the arcing that occurs at the spark plug, there is also an arc inside the distributor cap between the cap and the rotor. All of the secondary ignition wiring has a high resistance to reduce the affect of this arcing.

The coil wire can suffer from several possible problems. As the coil wire ages its resistance tends to increase. Also, as the wire ages the insulating quality of the jacket decreases, which makes it possible for the high voltages being carried by the wire to penetrate the jacket and arc to ground. Corrosion can also affect the current carrying ability of the wire.

A defective coil wire can result in misfiring, no-start, and poor power.

Distributor Cap and Rotor

These two components operate as a team. The coil wire delivers the high voltage to the center terminal of the distributor cap. A carbon conductor carries the voltage to the center of the rotor. The rotor will have either a metal or carbon resistive conductor that carries the voltage to the tip of the rotor. The rotor mounts on the top of the distributor shaft and is driven by the camshaft. As the rotor rotates, it approaches either copper or aluminum conductors on the inside of the distributor cap and arcs to these conductors, which carry the voltage to the spark plug wires.

The distributor cap is prone to cracking, corrosion, and carbon tracking. Carbon tracking occurs when a microscopic crack or piece of dirt provides a current path to ground that is easier than the current path and plug wire. The rotor is subject to corrosion and perforation. Perforation occurs when the high voltage seeks and finds a ground through the rotor to the distributor shaft.

Routine replacement of the distributor cap and rotor can prevent unforeseen problems. It is not necessary to replace the cap and rotor at each tune-up as many professional technicians recommend, but they should be replaced at every other tune-up. As you read this, however, do not assume that the preceding statement means that your mechanic has been ripping you off for the past 10 years. There is no disservice in a mechanic charging you $30 or $40 extra at each tune-up to ensure that you have less chance of developing premature problems.

As discussed in Chapter 2, when replacing the distributor cap or rotor, I advise that you replace them as a set. I further advise that they be built by the same name brand manufacturer. I have had situations in the past where a mismatched cap and rotor caused the rotor air gap to be so large that the engine either failed to start or misfired.

Spark Plug Wires

Back in the late sixties and the early seventies, when I first got into the car repair business, we checked spark plug wires by starting the engine in a darkened shop. If there were sparks flying around under the hood, it meant that one or more of the plug wires had perforated and was arcing to ground. For several years in the mid-seventies, I worked almost exclusively on fuel injected imports. Upon returning to working on domestics in 1980, I was mystified at the poor quality of the plug wires. I remembered few problems in the seventies, now there were many problems.

The difference was not the quality of the wires, but rather the leaner air/fuel ratios demanded in the early eighties (and today). In the early seventies, we unwittingly compensated for defective spark plug wires by enriching the idle mixture screws on the carburetors. This option was not available on the sealed carburetors of the late seventies. Therefore, technicians were forced to replace defective spark plug wires rather than simply masking the problem with a richer mixture.

When replacing spark plug wires, the old adage, "You get what you pay for" is especially true. A $50 set of spark plug wires can easily outlast four $12 sets of wires. A marginal plug wire can cause a misfire when the engine is under an extreme load and on a modern fuel injected car can make the engine run rich.

Spark Plugs

Every tune-up includes replacing the spark plugs. There are many brands of spark plugs on the market, some good, some bad. Asking for opinions on which is the best brand of spark plug is like asking a group which is the best soft drink, it is largely a matter of personal opinion. What I have always done, when I had a choice, was to use the brand that the manufacturer installed at the factory. My thinking is that the manufacturer has a vested interest in choosing the spark plug that would provide the best driveability and has the least chance of requiring replacement within 12,000 miles. This method has rarely failed to provide either myself or my customers with good service.

The spark plug consists of a pair of electrodes separated by an air gap of between 0.028 and 0.075in. As the spark from the coil travels down the plug wire seeking ground, it must arc across this air gap. If this arc is exposed to a properly preheated, well atomized mixture of fuel and air, this spark will ignite the fuel.

As the spark plugs arc at a high frequency in high temperatures, the electrode material will slowly vaporize. This causes the gap to widen. The wider the gap, the higher the voltage required to initiate the spark across the gap. Eventually, the voltage required to initiate the spark across the gap will be greater than the coil is capable of generating and a misfire will occur.

Common problems associated with the spark plugs are misfiring and difficulty in starting.

Timing Control
Centrifugal Advance

A set of weights and springs advances the timing in response to changes in rpm. The design differs little from the design in a standard point/condenser ignition system. Defective or frozen weights can result in low power.

Vacuum Advance

Timing control in response to changes in engine load is accomplished through a standard vacuum advance unit. When the load on the engine is low, the vacuum is high; the vacuum advance pulls the pickup coil to the fully advanced position. When the engine comes under a load, the vacuum drops and the timing retards. If the timing fails to retard properly when the engine load increases, the engine will likely detonate.

Setting the Timing

When setting the ignition timing, it is best to follow the instructions outlined on the EPA sticker under the hood. If the instructions are missing or obliterated, disconnect the vacuum advance and adjust to the specification listed in this book.

Troubleshooting
No Start: Basic Tests

Always begin testing a no-start condition by testing the battery. Use a hydrometer to test the state of charge of the battery. Each cell should read 1.220 or higher at 80deg F.

Engine Does Not Crank

Use a test light to check for power at the red wire on the starter relay. If there is no power, inspect and repair the wiring between the battery and the starter relay. If there is power, connect the test light to the brown wire at the starter solenoid. Rotate the key to attempt to start the engine. If the test light glows, replace the starter. If the test light does not glow, connect it to the yellow wire at the starter relay and try to start the engine. If the test light glows, connect the test light to the brown wire at the relay and try to start the engine. If the test light does not glow, replace the starter relay.

Testing the Distributor Pickup

If the engine cranks but does not start, remove the distributor cap. You will find two small terminals on the ignition module where the pickup coil connects. Disconnect these terminals from the module. The resistance between these wires is the resistance of the pickup coil. If the resistance is greater than 1120 or less than 920, replace the pickup coil.

Testing the Module

To test the ignition module, remove it from the distributor. Hold the ignition module right side up with the terminals toward you. Connect a test light to right terminal. This is the output terminal. Connect the other end of the test light to the positive terminal of the battery. Also connect the positive terminal of the battery to the large terminal on the left of the ignition module. Connect the negative terminal of the battery to the case or metal backing plate of the module. Using an ohmmeter as the low voltage power source, connect the ohmmeter leads across the smaller terminals on the module. The test light should light when the ohmmeter is connected and go out when the test light is disconnected. Replace the ignition module if the test light does not go on and off properly.

Starts But Does Not Continue to Run When the Key Is Released

Since this system uses no ballast resistor, the most likely cause of this symptom is a defective ignition switch. Be sure the spark plugs are in good condition before replacing the switch. Occasionally, a set of spark plugs will allow the engine to start but will not supply an intense enough spark to keep the engine running. When the engine is started, the typical owner of a carbureted car will pump the accelerator several times prior to turning the key. This provides the combustion chambers with an extremely rich mixture to start the engine. In this rich environment, the spark plugs have an easy task of igniting the mixture. As soon as the engine starts the mixture leans out and the engine dies.

If the spark plugs are new, or in good condition, connect a test light to the black/white wire at the ignition module and start the engine. If the test light goes out as the engine dies, inspect the wires and connections from the ignition switch to the coil and ignition module carefully. If the wires and connections are in good condition, replace the ignition switch. If the test light remains on while the engine dies, replace the ignition module.

Dies While Driving Down the Road

Testing the distributor pickup: Intermittent failure such as this is always frustrating. One of the imps of ancient Greek mythology is assigned to the creation of intermittent car problems. As a sentient being, this imp knows when a technician or car owner is looking for the source of the intermittent fault and allows the car to perform perfectly. As soon as the car gets into heavy, freeway traffic, the imp allows the fault to return.

While a distributor pickup can be easily tested for a no-start condition, testing it for an intermittent failure is virtually impossible. There are no valid resistance checks; the only way to test a suspected distributor pickup is to replace it and see if the symptom disappears. Yes, as you have already recognized, this is the classic "replace with known good unit" diagnostic technique. To add to the reader's frustration, this is exactly the same procedure that must be used to troubleshoot the ignition module for intermittents.

Before replacing either the ignition module or the distributor pickup, it must be confirmed that several other components are in good condition. Confirm the ignition switch is in good condition using the technique discussed above. Check the wires and connections to the ignition module, distributor, and coil. If they are in good condition, connect a test light to the number 2 terminal of the ignition module. Position the test light where you can see it as you drive down the road. When the car dies, is the test light on? If so, replace either the distributor pickup or the ignition module. If the light goes out, repair the intermittent open, short, or ground.

If the wiring test results indicate the problem is not in the wiring, you must make a decision. The problem is either in the ignition module, distributor pickup, or wiring between them. Inspect the wiring carefully. If the wiring is in good condition, call a few parts houses. The price of the ignition module and the distributor pickup is usually about the same, within ten dollars of one another. Since the ignition module is much easier to replace, change it first. If the problem persists, replace the distributor pickup.

Misfire at Idle

This is almost always never the fault of the distributor pickup ignition system; it is almost always the fault of some component in the secondary side of the ignition system.

Before troubleshooting any misfire, it is essential to verify that the engine is in good condition. A compression test is a good starting point. If the valves are adjustable, be sure that they are properly adjusted.

With a pair of "sissy" pliers, remove and replace one plug wire at a time from the spark plugs. As each plug wire is removed, the engine rpm should drop. If one of the cylinders fails to produce as great a drop in rpm as the others, that cylinder is the source of the misfire.

Assuming the cylinder is in good condition and the valves are properly adjusted, remove the spark plug wire for that cylinder and check the resistance. The resistance should be less than 10,000 ohms per volt.

If the resistance is correct, replace the spark plug. Unless the spark plugs are very new, replace them all.

Misfire Under a Load

Assuming the engine is in good condition, begin troubleshooting this problem by checking the spark plug gap. If they are gapped properly, replace the spark plugs. Even new spark plugs can misfire under a load.

If replacing the spark plugs does not solve the problem, remove the distributor cap. Inspect the wiring to the points. Frayed wiring can cause an intermittent open circuit as the vacuum advance moves the breaker plate. The intermittent open can cause a misfire.

Lack of Power

There are many things that can cause a lack of power, some related to the ignition system, some not. Begin checking this problem by confirming the engine is in good condition as are the air and fuel filters.

If a lack of power is the result of problems in the ignition system, it is likely the problem is in the timing control system. To test the timing control system, connect a timing light to the engine. Disconnect the vacuum advance and plug in the hose. With the engine at idle speed, check the timing. Now raise the engine speed to 2000 to 2500rpm. If the timing does not advance, the centrifugal advance system is not working. Inspect the distributor weights. If they are free and move easily, replace the weight springs. If the springs are weak, they will allow the timing to advance all the way prematurely, even at idle. If the weights are frozen, use penetrating oil or whatever is necessary to free them. If they are badly corroded, it may be necessary to replace the distributor.

If, or when, the centrifugal advance is working properly, with the engine still at 2000 to 2500rpm, reconnect the vacuum hose to the vacuum advance. When the vacuum hose is reconnected, the timing should advance several degrees.

Electronic Lean Burn 1976-1977

10

In the history of the world, three absolutely terrifying events have occurred: the eruption of Mount Vesuvius in AD 70, the Black Plague of Europe during the 1340s and 1350s, and the introduction of Chrysler's Lean Burn in 1976. All joking aside, the concept of Lean Burn is sound, the execution of the technology in 1976 was premature.

Lean Burn was an effort to control the production of oxides of nitrogen (NO_x) without the use of an expensive three-way catalytic converter. When the air/fuel ratio is burned leaner than 18:1, little NO_x is produced. When the air/fuel ratio is this lean, the mixture is difficult to ignite in the combustion chambers. As a result, the ignition timing must be kept as advanced as possible under all driving conditions. This early version of the Lean Burn system began with the timing at initial when the engine was idling and continued at idle as the engine speed increased. This made the car sluggish on start-up or after an extended idle.

As the engine speed continued to remain high, the computer, which was mounted on the air cleaner, would advance the timing gradually to a maximum of well over 30deg. When the throttle was closed, the timing would immediately drop back to initial. As soon as the throttle was opened again, the timing would jump back to where it was before the throttle was closed. The idea of this was to insure there was plenty of power as the throttle was opened and closed during cruise and coast. The problem was the computer could not tell the difference between coast and a short idle. After coming to a brief stop at a stop light or sign, the engine would often detonate violently as the timing advanced too much too fast.

Primary Ignition Components
Electronic Control Unit

The ECU mounts on the air cleaner on all but some police applications. At first it may seem odd to place temperature sensitive electronic components on top of the air cleaner. Actually, the fresh air being pulled into the air cleaner through the fresh air tube served to keep the module cool. One of the negative rumors about the Lean Burn system concerned its tendency to shut down intermittently. This condition could be caused by overheating the module. There was a tendency among the mechanics of the day to pay little or no attention to the fresh air tube that ran from the air cleaner to the radiator bulkhead. On this system, it was essential that this tube be in place and in good condition.

The ECU contained two separate modules, one to control the ignition timing and the other to control the firing of the coil. There are twelve wires connected to the ECU. There is an eight pin connector and a four pin connector.

On the eight pin connector,

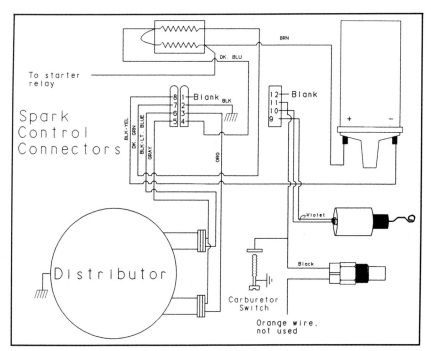

The Lean Burn ignition system may very well be the most slandered ignition system in history. The system was introduced as an effort to allow combustion in an engine where the air/fuel ratio was set at 18:1. This was beyond the capability of the spark plugs and other secondary ignition components of the seventies. Instead of putting the blame where it truly belonged, the Lean Burn system was faulted.

The Lean Burn system of 1978 used a single pickup coil for both starting and running the engine.

All models of the Lean Burn system, except those built in 1978, used one pickup coil for starting and a second for running. In the event of a failure in the run pickup coil, the start pickup coil will take over and limp the vehicle home.

terminal 1 is connected to nothing. Terminal 2 connects the ECU to ground through a black wire. An orange wire connects terminal 3 to the run pickup coil. Terminal 4 has a dark blue wire that carries power to the ECU. The other side of the run pickup coil is connected to terminal 5 through a gray wire. This wire is also the common wire for the start pickup coil. The black and light blue wire that connects the distributor to the ECU is the other wire for the start pickup coil. This wire is connected to terminal 6 of the ECU. Terminal 7 (a dark green wire) carries reduced current and voltage from the outbound side of the ignition coil to the ECU. The black and yellow wire connected to terminal 8 runs to the negative terminal of the coil. This is the wire that the ECU uses to fire the ignition coil.

The four pin connector is numbered 9 through 12. Pins 9 and 10 go to the throttle position transducer. These wires are violet. A black wire runs from terminal 11 to both the coolant sensor and the throttle switch. Terminal 12 is connected to nothing.

Pickup Coil

Both of the pickup coils are reluctance pickups. These sensors produce an alternating current as the distributor rotates. The sensors consist of a coil of wire, permanent magnet, and reluctor wheel. The reluctor wheel is part of the distributor shaft. As the reluctor wheel rotates, it distorts the magnetic field back and forth across the coil of wire. This distorting induces an alternating current in the coil of wire. As the speed of distributor shaft rotation increases, the amplitude (voltage output) of the sensor increases. More important, since the AC signal is a sine wave, the frequency of the sine wave changes as the speed of the distributor shaft changes. The frequency of the sine wave tells the ignition module the speed and position of the distributor shaft. Since the distributor shaft is synchronized to the cam shaft, the ignition is synchronized to the mechanical components of the engine. The ECU uses the signal from each of the pickup coils as a trigger source for the firing of the ignition coil. One of the pickups is used while the engine is running; the other is used only for starting and as a backup should the run pickup fail.

Ignition Switch

The ignition switch controls whether current will be available to the coil and ignition module.

Throttle Position Transducer

This sensor informs the ECU of the throttle's movement. Primarily, the ECU is concerned about the rate of throttle change.

Coolant Sensor

The coolant sensor is a switch that closes when the temperature of the engine drops below 150deg F. Although there are two wires connected to the coolant sensor, only the black has a function. The orange wire has no purpose except to confuse unwary technicians. The coolant switch input to the ECU allows full timing advance anytime the throttle is not closed and the temperature is below 150deg. When the temperature rises above 150deg F, the tendency to detonate increases; and the signal from the coolant sensor tells the ECU to limit the amount of allowable advance.

Carb Switch

The carb switch informs the computer when the throttle is closed. The throttle switch grounds an electrical signal that originates at terminal 11 of the ECU. This is the same terminal that is connected to the coolant temperature switch. If the temperature of the engine is below 150deg F and the throttle is open, maximum timing advance

In place of the vacuum advance unit, the Lean Burn system uses a vacuum transducer to inform a computer about the load on the engine. The computer *then uses that information in choosing how much timing advance to allow.*

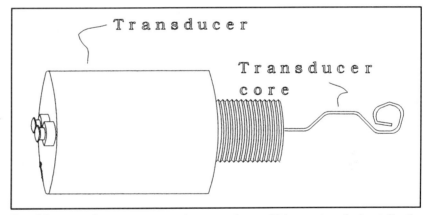

In addition to the vacuum transducer, most models of the Lean Burn system also use a throttle position trans- *ducer. This sensing device tells the computer what the driver expects from the engine.*

is allowed. When the throttle closes on the cold engine, the timing is immediately cut back to the level it would be at if the engine were warm.

Vacuum Transducer

Mounted on the side of the ECU is a round sensor with a vacuum hose attached. We have already established in earlier chapters that the two major factors in the control of ignition timing are rpm and engine load. As the speed of the Lean Burn engine increases, the ECU sees an increase in the frequency of the pulse coming from the distributor pickup coil. As the frequency increases, the ECU allows an increased limit on the timing advance. The vacuum transducer monitors the load on the engine. As engine load increases, the vacuum in the intake manifold decreases. The vacuum transducer informs the ECU of the decreasing vacuum and the timing is retarded.

The Lean Burn system was introduced in 1976. At the time there was not a mechanic in the aftermarket sector that was freely willing to even screw in a set of spark plugs. The system was, to say the least, poorly understood. The aftermarket trainers that gave classes on the topic treated it as a system so complex that it was beyond the ability of most people's understanding. At the time, many technicians continued to take the same shortcuts when doing a tune-up that they had been able to get away with for decades. The Lean Burn system did not permit these shortcuts.

Because of the sophistication of the electronics, the automotive technician, the tune-up technician to be more precise, who had always been allowed a relatively large margin for error, was suddenly faced with the necessity for precision. (Note that the word "sophistication" is derived from the ancient Greek "sophos" which meant "wise or clear" and is further derived from the Latin "sophisticus" which meant "naive." The author is not sure which derivation is most appropriate.)

The carburetor used with this system was adjusted to an air/fuel ratio of 18:1. This air/fuel ratio was lean to the point of almost being noncombustible; the slightest vacuum leak would render the mixture noncombustible. Prior to Lean Burn, it had been possible to cover up vacuum leaks by simply enriching the mixture. The Lean Burn carburetor would not permit this. This lean mixture also meant that it was necessary to maximize the amount of energy available in the combustion chambers to ignite the air/fuel mixture. It was essential that the entire secondary ignition be in good condition, including the spark plugs, distributor cap and rotor, and plug wires.

Secondary Ignition Components
Coil

The ignition coil secondary consists of hundreds of windings of very thin wire. When current stops flowing through the primary, the magnetic field created by the primary current flow collapses. This collapsing magnetic field induces several thousand volts in the secondary. This is the voltage that is used to jump the gap of the spark plugs to fire the mixture in the cylinders.

Because of the relatively low current in this high-voltage side of the coil, there are relatively few problems in the secondary side of the coil. The problems that do occur are usually in the form of open circuits.

Coil Wire

The coil secondary output wire carries the high-voltage current from the coil to the distributor cap. The coil wire is normally 6 to 12in long and has a resistance of a few thousand ohms. This relatively high resistance helps to reduce the intensity of the radio signal created by the secondary ignition.

All arcing generates a radio signal. In addition to the arcing that occurs at the spark plug, there is also an arc inside the distributor cap between the cap and the rotor. All of the secondary ignition wiring has a high resistance to reduce the affect of this arcing.

The coil wire can suffer from several possible problems. As the coil wire ages its resistance tends to increase. Also, as the wire ages the insulating quality of the jacket decreases, which makes it possible for the high voltages being carried by the wire to penetrate the jacket and arc to ground. Corrosion can also affect the current carrying ability of the wire.

A defective coil wire can result in misfiring, no-start, and poor power.

Distributor Cap and Rotor

These two components operate as a team. The coil wire delivers the high voltage to the center terminal of the distributor cap. A carbon conductor carries the voltage to the center of the rotor. The rotor will have either a metal or carbon resistive conductor that carries the voltage to the tip of the rotor. The rotor mounts on the top of the distributor shaft and is driven by the camshaft. As the rotor rotates, it approaches either copper or aluminum conductors on the inside of the distributor cap and arcs to these conductors, which carry the voltage to the spark plug wires.

The distributor cap is prone to cracking, corrosion, and carbon tracking. Carbon tracking occurs when a microscopic crack or piece of dirt provides a current path to ground that is easier than the current path and plug wire. The rotor is subject to corrosion and perforation. Perforation occurs when the high voltage seeks and finds a ground through the rotor to the distributor shaft.

Routine replacement of the distributor cap and rotor can prevent unforeseen problems. It is not necessary to replace the cap and rotor at each tune-up as many professional technicians recommend, but they should be replaced at every other tune-up. As you read this, however, do not assume that the preceding statement means that your mechanic has been ripping you off for the past 10 years. There is no disservice in a mechanic charging you $30 or $40 extra at each tune-up to ensure that you have less chance of developing premature problems.

As discussed in Chapter 2, when replacing the distributor cap or rotor, I advise that you replace them as a set. I further advise that they be built by the same name brand manufacturer.

I have had situations in the past where a mismatched cap and rotor caused the rotor air gap to be so large that the engine either failed to start or misfired.

Spark Plug Wires

Back in the late sixties and the early seventies, when I first got into the car repair business, we checked spark plug wires by starting the engine in a darkened shop. If there were sparks flying around under the hood, it meant that one or more of the plug wires had perforated and was arcing to ground. For several years in the mid-seventies, I worked almost exclusively on fuel injected imports. Upon returning to working on domestics in 1980 I was mystified at the poor quality of the plug wires. I remembered few problems in the seventies, now there were many problems.

The difference was not the quality of the wires, but rather the leaner air/fuel ratios demanded in the early eighties (and today). In the early seventies, we unwittingly compensated for defective spark plug wires by enriching the idle mixture screws on the carburetors. This option was not available on the sealed carburetors of the late seventies. Therefore, technicians were forced to replace defective spark plug wires rather than simply masking the problem with a richer mixture.

When replacing spark plug wires, the old adage, "You get what you pay for" is especially true. A $50 set of spark plug wires can easily outlast four $12 sets of wires. A marginal plug wire can cause a misfire when the engine is under an extreme load and on a modern fuel injected car can make the engine run rich.

Spark Plugs

Every tune-up includes replacing the spark plugs. There are many brands of spark plugs on the market, some good, some bad. Asking for opinions on which is the best brand of spark plug is like asking a group which is the best soft drink, it is largely a matter of personal opinion. What I have always done, when I had a choice, was to use the brand that the manufacturer installed at the factory. My thinking is that the manufacturer has a vested interest in choosing the spark plug that would provide the best driveability and has the least chance of requiring replacement within 12,000 miles. This method has rarely failed to provide either myself or my customers with good service.

The spark plug consists of a pair of electrodes separated by an air gap of between 0.028 and 0.075in. As the spark from the coil travels down the plug wire seeking ground, it must arc across this air gap. If this arc is exposed to a properly preheated, well atomized mixture of fuel and air, this spark will ignite the fuel.

As the spark plugs arc at a high frequency in high temperatures, the electrode material will slowly vaporize. This causes the gap to widen. The wider the gap, the higher the voltage required to initiate the spark across the gap. Eventually, the voltage required to initiate the spark across the gap will be greater than the coil is capable of generating and a misfire will occur.

Common problems associated with the spark plugs are misfiring and difficulty in starting.

Timing Control
Centrifugal Advance

The Lean Burn System uses a computer to control the ignition timing. As the pulse frequency from the pickup coil increases, the ECU advances the timing.

Vacuum Advance

There is no mechanical movement associated with the advancing of the ignition timing as the engine load decreases. Instead, the vacuum transducer located on the computer monitors engine load and signals the computer to retard the timing as the load increases and to retard the timing as the load decreases.

Setting the Timing

To adjust the timing in the Lean Burn system, read the sticker found under the hood and follow the printed instructions precisely. There are several variations on the adjustment procedure depending on the year and model of the vehicle. If the sticker is missing, refer to the factory service manual. The manual will describe a procedure similar to the following.

1. Check the air gap between the pickup and the reluctor wheel with a brass feeler gauge. This gap should be 0.008in. A gap that is incorrect can affect the timing. This is one of the essential tune-up steps that is almost always skipped. (Please note that the factory service manual may assume you have done this prior to consulting the manual and not mention the adjustment.)

2. For most early Lean Burn applications, the timing setting is given for the engine idling in drive. The safest way to check or set the timing specification is to ask a trustworthy person to sit in the car and hit the brake if it should suddenly move forward. (I recommended that you not select estranged spouses or strange neighbors for this job.) The brake pedal should not be depressed during the adjustment sequence. The power brake booster can affect engine vacuum, this can result in inaccurate settings.

Apply the parking brake. Test the brake carefully to ensure it will hold the car when applied. Block the wheels and ask the selected person to sit in the driver's seat, just in case. Connect an advance timing light and place a jumper wire between the electrical contact on the carburetor body and ground. This electrical contact is the carb switch. Start the engine and allow it to idle for several minutes. When the engine is warm, raise the rpm above 1500 momentarily then allow it to idle for about two minutes.

3. Adjust the engine idle speed to *exactly* what is specified on the sticker found under the hood or the factory service manual. An error of even 100rpm or less can cause serious driveability problems. This specification is usually given as an idle speed in drive setting. Set the timing as specified on the sticker found under the hood or in the factory service manual.

Troubleshooting
No Start: Basic Tests

Ask anyone with professional experience on the early Lean Burn systems and they are likely to tell you that if the engine does not start, replace the computer. This is expensive and often incorrect.

1. Always begin testing a no-start condition by testing the battery. Use a hydrometer to test the state of the battery's charge. Each cell should read 1.220 or higher at 80deg F. Remember that the air/fuel mixture being delivered by the carburetor to the cylinders is extremely lean. A lean mixture is difficult to ignite. If the electrical power available to the ignition coil is low, it may not create a powerful enough spark to ignite the mixture. This is most critical during start-up because the starter is drawing large amounts of energy from the battery.

2. Disconnect the coolant switch and place a piece of paper between the carb switch and throttle screw.

3. Connect a test light to the negative terminal of the ignition coil. Crank the engine and observe the test light. If the light blinks on and off, the problem is in the secondary side of the system. Inspect the cap, rotor, plug wires, and spark plugs. With the Lean Burn system, it is a mistake to assume that any spark at the spark plug is a sufficient spark to ignite the extremely lean mixture in the engine. Any secondary ignition parts that are even marginal should be replaced.

4. If the test light did not blink, check the ballast resistor. The resistance of the resistor on one side should be 0.5 to 1.25 ohms. The resistance of the resistor on the other side should be 4.75 to 5.75 ohms. If the ballast resistor is good, connect the red lead of a voltmeter to the carb switch terminal. Connect the black lead to a good engine ground. Turn the ignition switch to the run position and read the voltmeter. If the voltmeter reads less than 5 volts, turn the ignition switch off, disconnect the "double" terminal on the underside of the Spark Control Computer. Turn the ignition switch back to the run position. The voltage at terminal 4 should be battery voltage. If the voltage is less, check the wiring from terminal 4 to the ignition switch for opens or grounds.

There are several other circuits that are parallel to this terminal off the ignition switch. It is the main power source for circuits such as the radio, rear window defogger (Chrysler calls this the Electric Back Light), and other devices that only operate while the key is in the run position. A grounded wire in one of these parallel circuits can cause a no-start condition.

If the voltage at terminal 4 was battery voltage, turn the ignition switch off and disconnect the single connector at the Spark Control Computer. With an ohmmeter, check for continuity between terminal 11 and the carb switch. The resistance should be very close to zero. If the resistance is higher, repair the wire. If there is continuity through the wires, connect one lead of the ohmmeter to engine ground. The resistance between either end of the wire that runs between terminal 11 and the carb switch should be infinity. If it is less than infinity, the wire is grounded and must be repaired.

There should be no resistance between terminal 2 and ground. This wire is the main ground for the computer. The resistance between terminals 5 and 6 as well as 5 and 3 should be 150 to 900 ohms. These are the wires that carry the signal from the pickup coils to the Spark Control Computer. If the resistance is outside this specification, repair the wires as needed. If the wires are in good condition, replace the offending pickup coil.

Testing the Module

If the system passes all of the above tests and the test light fails to blink when connected to the negative side of the ignition coil while the engine is being cranked, replace the Spark Control Computer.

Starts But Does Not Continue to Run When the Key Is Released

The most likely cause of this problem is a defective ballast resistor. On rare occasions, the cause will be a defective run pickup coil. In most cases, the engine will continue to run off of the start pickup coil if the run pickup coil were defective; however, performance would be impaired. It is also possible that defective or marginal components in the ignition secondary can cause this symptom. While the engine is being cranked, the choke is fully applied and the mixture is extremely rich. When the engine starts, the vacuum break partially pulls the choke off. When the choke pulls off, the mixture leans out; the mixture becomes harder to ignite and the engine dies.

Dies While Driving Down the Road

Inspect the cap, rotor, plug

wires, and spark plugs. As stated above in number 3 of the "No Start: Basic Tests" section, with the Lean Burn system, it is a mistake to assume that any spark at the spark plug is a sufficient spark to ignite the extremely lean mixture in the engine. Any secondary ignition parts that are even marginal should be replaced. If either the pickup coil or ballast resistor run circuits were to fail, the ignition system would theoretically revert to the start pickup coil and ballast resistor circuit. This is supposed to provide a backup or limp-in mode of operation. In this limp-in mode the timing does not advance or retard. With the timing fixed, driveability suffers. The driveability problem can cause the engine to stall, particularly on deceleration.

To test the ballast resistor, replace it. This may sound like a diagnostician's cop-out, but the fact is that a bad ballast resistor will often test good when it is cold, yet it will become an open circuit when heated to operating temperature. The typical person will spend more on gasoline getting to the parts store than he will pay for the ballast resistor.

If the problem persists after the ballast resistor is replaced, connect an ohmmeter between terminal 5 (the gray wire) and terminal 3 (the orange wire) of the eight terminal connector on the Electronic Spark Control (ESC) computer. The resistance should be between 150 and 900 ohms. If the resistance is within these specifications, heat the distributor pickup coil with a hair dryer set on the hot setting or with a drop-light. When the pickup is as hot as it is going to get, tap it with the butt end of a screwdriver. If the resistance reading varies at any point during the above test, replace it. If the resistance does not fluctuate, replace the ESC computer. If replacing the ballast resistor and the run pickup does not eliminate the problem, replace the ESC computer.

Misfire at Idle

This is almost always never the fault of the Lean Burn system; it is almost always the fault of some component in the secondary side of the ignition system.

Before troubleshooting any misfire, it is essential to verify that the engine is in good condition. A compression test is a good starting point. If the valves are adjustable, be sure that they are properly adjusted.

With a pair of "sissy" pliers, remove and replace one plug wire at a time from the spark plugs. As each plug wire is removed, the engine rpm should drop. If one of the cylinders fails to produce as great a drop in rpm as the others, that cylinder is the source of the misfire.

Assuming the cylinder is in good condition and the valves are properly adjusted, remove the spark plug wire for that cylinder and check the resistance. The resistance should be less than 10,000 ohms per volt. If the resistance is correct, replace the spark plug. Unless the spark plugs are very new, replace them all.

Misfire Under a Load

Assuming the engine is in good condition, begin troubleshooting this problem by checking the spark plug gap. If they are gapped properly, replace the spark plugs. Even new spark plugs can misfire under a load.

If replacing the spark plugs does not solve the problem, remove the distributor cap. Inspect the wiring to the points. Frayed wiring can cause an intermittent open circuit as the vacuum advance moves the breaker plate. The intermittent open can cause a misfire.

Lack of Power

Inspect the cap, rotor, plug wires, and spark plugs. Again, with the Lean Burn system, it is a mistake to assume that any spark at the spark plug is a sufficient spark to ignite the extremely lean mixture in the engine. Any secondary ignition parts that are even marginal should be replaced.

Confirm the problem is in the timing control system by performing the following test. With the ignition switch turned off, connect a timing advance timing light. Insulate the carb switch with a piece of paper. This will trick the ESC computer into believing the throttle is open and the driver is trying to accelerate. Connect a hand-held vacuum pump to the vacuum transducer and apply 16in of vacuum. Ensure the vacuum pump holds the 16in throughout the next ten minutes. Start the engine, sit back, and read the sports section of the local newspaper.

After ten minutes, the timing should have advanced to between 18 and 35deg. Unfortunately, it is necessary to refer to the latest service manual and service bulletins to get the latest opinion on the proper specification. If you do not have access to these specifications, anything in the 18 to 35deg range can be considered acceptable. If the timing fails to advance properly, the lack of power problem is related to the ESC computer problem and you should proceed with the following tests. If the timing does advance properly, the lack of power is related to engine condition, secondary ignition problems, or transmission problems.

Since the timing control system is totally electronic in the Lean Burn System, there are no distributor weights or vacuum advance units. The advance function that results when the speed of the engine increases is based on the signal from the run pickup coil. If the run pickup coil were to fail, the ESC computer would not know if the speed of the engine were increasing and would not advance the timing.

To test the run pickup coil, connect an ohmmeter between terminal 5 (the gray wire) and terminal 3 (the orange wire) of the eight terminal connector on the ESC computer. The resistance should be between 150 and 900 ohms. If the resistance is within these specifications, heat the distributor pickup coil with a hair dryer set on the hot setting or with a drop-light. When the pickup is as hot as it is going to get, tap it with the butt end of a screwdriver. if the resistance reading varies at any point during the above test, replace it.

Two sensors, the coolant temperature sensor and the throttle position transducer, can affect the operation of the ESC computer. Testing the coolant sensor is easy. When the temperature of the sensor is less than 150deg F (66deg C), the resistance between the terminals should be almost zero. When the temperature rises above 150deg F, the switch inside the sensor should be open.

Test the resistance through the throttle position transducer. The resistance should be 50 to 90 ohms. Connect an ohmmeter between terminals 9 and 10 on the ESC computer. If the resistance is outside the specifications, check the violet wires to the throttle position transducer for opens, short, and grounds. If the wires are in good condition, replace the transducer.

If the pickup coil, coolant sensor, and throttle transducer test good, the number of possibilities is suddenly and dramatically reduced. Inspect the hose from the intake manifold to the vacuum transducer on the ESC computer for damage and restrictions. If the hose is in good condition, replace the ESC computer.

Later Lean Burn Systems

Although the classic (a.k.a. troublesome) Lean Burn system was discontinued in 1977, variations on the system were used until 1990. These variations include systems used on light-duty trucks and in particular on the non-fuel injection applications. Only one model of fuel injected car, the 1981 through 1983 Chrysler Imperial, borrowed from Lean Burn technology. The ways that these subsequent systems differ are at the same time minimal and profound. To the engineer, the computer changed from analog to the more dependable digital technology. Some models eliminate the throttle position transducer, some retain it. Some models use a single pickup coil, some continue to use the same dual pickup technology used in the 1976-1977 models. Some replace the reluctance pickup coil with a Hall Effects.

Chrysler EFI and Optical Ignition   *11*

Chrysler began its emphasis on the use of electronic fuel injection in 1984. Its ill-fated attempt to use fuel injection on the 1981-1983 Imperial led to the development and use of a modern injection system beginning in 1984. By 1988, there were almost no Chrysler applications that used a carburetor.

Primary Ignition Components

Most fuel-injected Chrysler applications use a Hall Effects distributor. Its primary advantage over the VRT is its ability to detect position and rotational speed from zero rpm to tens of thousands. Its primary disadvantage is that it is not as rugged as the VRT and is more sensitive to errant magnetic fields. An intense magnetic field can shut down the proper operation of a Hall Effects.

A Hall Effects pickup is a semiconductor carrying a current flow. When a magnetic field falls perpendicular to the direction of that current flow, part of that current is redirected perpendicular to the main current path. The semiconductor is placed near a permanent magnet. A set of metal blades, or armature, attached to a rotating shaft or other device passes between the Hall Effects semiconductor and the permanent magnet. As the armature rotates, the magnet field is alternately applied to the Hall Effects and interrupted. The result is a pulsing current perpendicular to the main current path. This frequency is directly proportional to the speed of armature rotation. Since the output is only dependent on the presence of the magnetic field, the Hall Effects unit is capable of detecting armature position even when there is no rotational speed.

The turbocharged Chrysler

All fuel injected applications except the 3.0 liter use a Hall Effects pickup to sense distributor position and speed.

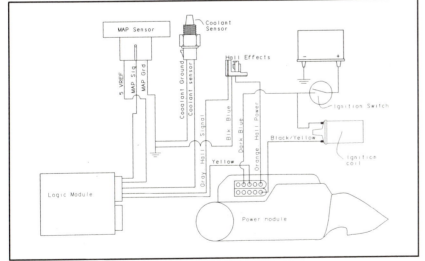

In late-model Chrysler applications, the fuel injection computer is also responsible for controlling ignition timing. Using information from the manifold absolute pressure (MAP) sensor from the coolant temperature sensor and from the Hall Effects in the computer, the computer decides how much to advance the timing. From 1984 through 1987, the fuel injection computer had two separate sections. The Logic Module gathered information about the operating conditions of the engine, made decisions, then sent its decision off to the Power Module, which made sure these decisions were carried out.

The turbocharged applications have a double Hall Effects pickup. Even though this is from a non-turbocharged application, the mounting for the second magnet can still be seen.

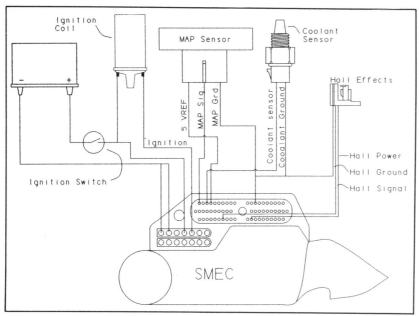

Beginning in 1987, the Logic Module and Power Module were combined into one unit called the Single Module Engine Controller (SMEC). While in 1987 the SMEC was only used on the 3.0 liter engine applications, by 1988 it was used on all fuel injection applications.

applications produced from 1984 through 1992 use a dual Hall Effects unit. The second Hall Effects pickup is used to trigger the injectors or injector groups.

In 1987, when Chrysler introduced the 3.0 liter engine in the Dodge Caravan, the company introduced a new ignition system. The most significant difference between this and other electronic ignition systems is the pickup. The reluctance pickup and Hall Effects are replaced by an optical sensor. The optical sensor consists of two light emitting diodes (LEDs) and two photodiodes. Together, one LED and one photodiode make up an optical pair. Rotating between each LED/photodiode pair is a disc. The disc has two sets of slots in it. There are six slots in the inner set and 345 slots in the outer set. The inner set of slots is used

Unlike the old Chrysler Hall Effects ignition system, the blades in the fuel-injected version do not hang from the rotor. The blades in the fuel-injected models sit in the bottom of the distributor where they are more stable.

The 3.0 liter Chrysler/Mitsubishi engine uses an optical distributor. The tiny slots mark off individual degrees of distributor rotation.

to synchronize the ignition and injection system. The signals from the outer set is used to assist the computer (Chrysler calls it a Single Module Engine Controller or SMEC) in the control of the ignition timing.

In this system, both the ignition module and the timing control module are part of the main fuel injection computer. For all models from 1984 through 1987, the computer was a dual module type. One module, the thinking or Logic Module, was located in the passengers kick panel and gathered information from all the sensors. Once the information was gathered, the Logic Module processed the information, made decisions, and sent the decisions to the Power Module.

Located behind the left headlight, the Power Module received the commands from the Logic Module and turned the actuating devices, including the coil, on and off. On models produced from 1987 1/2 (the 3.0 liter minivan only, the SMEC was not introduced for other applications until 1988) through 1990, this computer was known as an SMEC. On later models, the computer was called a Single Board Engine Controller (SBEC). The SMEC features a 60 pin terminal for the low current circuits, such as sensors, and a 14 pin connector for high current flow circuits, such as coil negative. With the introduction of the SBEC, the 14 pin connector is eliminated.

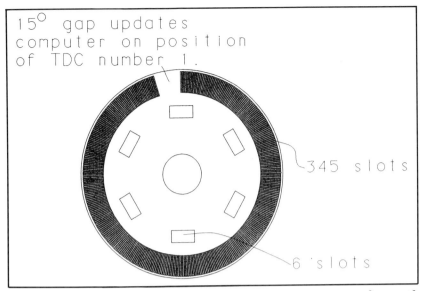

The optical disc used in the 3.0 liter has two sets of slots. The outer slots are used to control ignition timing.

The inner set of slots is used as a reference for triggering the fuel injectors.

There is an unusual configuration in the 3.0 distributor cap. In most caps, the electrode goes directly through the cap to the plug wire. In the 3.0 liter

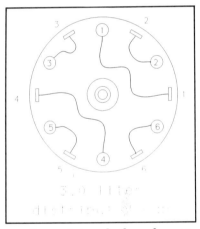

cap the spark travels through a conductor to the opposite side of the cap.

Secondary Ignition Components
Coil

The early model, through 1990, used the same oil filled coil that John Glenn had on the car he drove to Cape Canaveral in 1963. In later applications, the oil-filled coil is replaced by an E-core coil. The ignition coil secondary consists of hundreds of windings of very thin wire. When current stops flowing through the primary, the magnetic field created by the primary current flow collapses. This collapsing magnetic field induces several thousand volts in the secondary. This is the voltage that is used to jump the gap of the spark plugs to fire the mixture in the cylinders.

Because of the relatively low current in this high-voltage side of the coil, there are relatively few problems in the secondary side of the coil. The problems that do occur are usually in the form of open circuits.

Coil Wire

The coil secondary output wire carries the high-voltage current from the coil to the distributor cap. The coil wire is normally 6 to 12in long and has a resistance of a few thousand ohms. This relatively high resistance helps to reduce the intensity of the radio signal created by the secondary ignition.

All arcing generates a radio signal. In addition to the arcing that occurs at the spark plug,

83

The Logic Module is located inside the passenger's kick panel. Its job is to receive data from the various sensors connected to the engine and make decisions about fuel injector on-time and ignition timing. This is from a 1985 model, the only year the MAP sensor was mounted on the Logic Module.

The inside of the Logic Module is a complex maze of circuitry. Ultimately, this module—either directly or indirectly through the Power Module—controls everything going on under the hood.

The SMEC, located next to the battery, was introduced in 1987. Its location logically would expose the SMEC to a large amount of corrosion, yet corrosion problems are rare.

In the nineties, Chrysler replaced the oil-filled coils with the more modern solid coil.

Late-model Chrysler products have eliminated the oil-filled coil in favor of the solid core coil.

there is also an arc inside the distributor cap between the cap and the rotor. All of the secondary ignition wiring has a high resistance to reduce the affect of this arcing.

The coil wire can suffer from several possible problems. As the coil wire ages its resistance tends to increase. Also, as the wire ages the insulating quality of the jacket decreases, which makes it possible for the high voltages being carried by the wire to penetrate the jacket and arc to ground. Corrosion can also affect the current carrying ability of the wire.

A defective coil wire can result in misfiring, no-start, and poor power.

Distributor Cap and Rotor

These two components operate as a team. The coil wire delivers the high voltage to the center terminal of the distributor cap. A carbon conductor carries the voltage to the center of the rotor. The rotor will have either a metal or carbon resistive conductor that carries the voltage to the tip of the rotor. The rotor mounts on the top of the distributor shaft and is driven by the camshaft. As the rotor rotates, it approaches either copper or aluminum conductors on the inside of the distributor cap and arcs to these conductors, which carry the voltage to the spark plug wires.

The distributor cap is prone to cracking, corrosion, and carbon tracking. Carbon tracking occurs when a microscopic crack or piece of dirt provides a current path to ground that is easier than the current path and plug wire. The rotor is subject to corrosion and perforation. Perforation occurs when the high voltage seeks and finds a ground through the rotor to the distributor shaft.

Routine replacement of the distributor cap and rotor can prevent unforeseen problems. It is not necessary to replace the

The distributor on the four-cylinder Chrysler applications is hidden under a protective plastic cover. Care should be used when reinstalling this cover as the screws are easy to cross-thread.

The corrosion on the tip of the rotor does not indicate any specific problem. In fact the corrosion may have been painted on to the rotor during the manufacturing process. If you are in doubt about the condition of the rotor, replace it.

cap and rotor at each tune-up as many professional technicians recommend, but they should be replaced at every other tune-up. As you read this, however, do not assume that the preceding statement means that your mechanic has been ripping you off for the past 10 years. There is no disservice in a mechanic charging you $30 or $40 extra at each tune-up to ensure that you have less chance of developing premature problems.

As discussed in Chapter 2, when replacing the distributor cap or rotor, I advise that you replace them as a set. I further advise that they be built by the same name brand manufacturer. I have had situations in the past where a mismatched cap and rotor caused the rotor air gap to be so large that the engine either failed to start or misfired.

It should be noted that on most of the four cylinder applications, spark plug wires are locked into the distributor cap and can only be released by unlocking them from the inside of the cap. These metal locks also provide the secondary contacts for the rotor inside the cap. Furthermore, on the 3.0 liter cap, location of the plug wires does not correspond with the electrodes on the interior of the cap.

Spark Plug Wires

Back in the late sixties and the early seventies, when I first got into the car repair business, we checked spark plug wires by starting the engine in a darkened shop. If there were sparks flying around under the hood, it meant that one or more of the plug wires had perforated and was arcing to ground. For several years in the mid-seventies, I worked almost exclusively on fuel injected imports. Upon returning to working on domestics in 1980 I was mystified at the poor quality of the plug wires. I remembered few problems in the seventies, now there were many problems.

The difference was not the quality of the wires, but rather the leaner air/fuel ratios demanded in the early eighties (and today). In the early seventies, we unwittingly compensated for defective spark plug wires by enriching the idle mixture screws on the carburetors. This option was not available on the sealed carburetors of the late seventies. Therefore, technicians were forced to replace defective spark plug wires rather than simply masking the problem with a richer mixture.

When replacing spark plug wires, the old adage, "You get what you pay for" is especially true. A $50 set of spark plug wires can easily outlast four $12 sets of wires. A marginal plug wire can cause a misfire when the engine is under an extreme load and on a modern fuel injected car can make the engine run rich.

For those who wish to replace the spark plug wires on the Chrysler 4-cylinder applications for the first time, there is a surprise waiting...

...the plug wires clip into the distributor cap...

... and form the electrodes for the distributor cap. When the plug wires are removed from the cap, the cap has no electrodes.

Spark Plugs

Every tune-up includes replacing the spark plugs. There are many brands of spark plugs on the market, some good, some bad. Asking for opinions on which is the best brand of spark plug is like asking a group which is the best soft drink, it is largely a matter of personal opinion. What I have always done, when I had a choice, was to use the brand that the manufacturer installed at the factory. My thinking is that the manufacturer has a vested interest in choosing the spark plug that would provide the best driveability and has the least chance of requiring replacement within 12,000 miles. This method has rarely failed to provide either myself or my customers with good service.

The spark plug consists of a pair of electrodes separated by an air gap of between 0.028 and 0.075in. As the spark from the coil travels down the plug wire seeking ground, it must arc across this air gap. If this arc is exposed to a properly preheated, well atomized mixture of fuel and air, this spark will ignite the fuel.

As the spark plugs arc at a high frequency in high temperatures, the electrode material will slowly vaporize. This causes the gap to widen. The wider the gap, the higher the voltage required to initiate the spark across the gap. Eventually, the voltage required to initiate the spark across the gap will be greater than the coil is capable of generating and a misfire will occur.

Common problems associated with the spark plugs are misfiring and difficulty in starting.

Timing Control

For both the EFI and Optical ignition systems, the computer controls the timing. Three of the fuel injection sensors—the rpm, manifold absolute pressure (MAP), and coolant—serve double duty by also supplying information for the control of ignition timing.

The rpm sensor is the Opti-

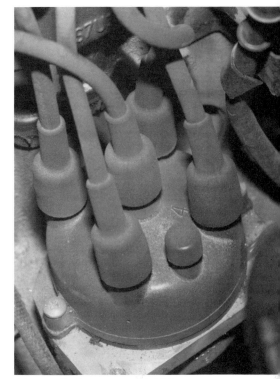

The spark plug wires should be routinely replaced at least during every other tune-up. Although the coil wire is easily removed, the spark plug wires must be unlocked from the inside of the cap.

Since the timing on EFI equipped Chryslers is adjusted by the fuel injection computer, the fuel injection sensors must be used to control the ignition timing. In these systems, the information about engine load originates with the MAP sensor.

The MAP sensor is usually mounted on the rear engine compartment bulkhead, the left shock tower or the right (passenger) side of the engine.

cal or Hall Effects sensor. In addition to supplying the signal to trigger the ignition primary and the injectors, it also replaced the weights of the old style ignition timing control systems.

Vacuum Advance is replaced by an input to the computer from the MAP sensor. This sensor consists of an extremely thin diaphragm strung between four variable resistors that form a Wheatstone Bridge. When the pressure on the diaphragm increases, the resistors are stretched causing their resistance to change. The result is that the output voltage of the sensor varies from reference voltage as the pressure on the diaphragm changes. For most applications using this type of sensor, full atmospheric pressure causes an output voltage of about 4.5 volts (where reference voltage is 5 volts). If the sensor is being used to measure manifold pressure, this voltage will drop to about 1.5 volts when the engine is idling. (These voltages are approximate.) This drop in voltage is proportional to the drop in manifold pressure that occurs when the engine is at an idle.

The following is an explanation of this concept.

Normal Barometric Pressure = 29.92in of mercury = 15psi = 100kPa [kilopascal]

Variable voltage MAP output = 4.5 volts

Manifold pressure at an idle = 10in of mercury= 5psi = 35kPa

Note that these pressures are about one-third of atmospheric.

MAP sensor output voltage = 1.5 volts

Turbocharged applications that use a variable voltage pressure sensor to measure manifold pressure simply cut the above voltage readings in half. This accommodates the accurate measurement of pressures greater than atmospheric when under boost.

The coolant sensor sends a signal to the Logic Module, SMEC, or SBEC. When the sensor tells the computer the engine is cold, the computer allows more timing advance. When the computer senses the engine warming, timing advance is limited to reduce the tendency of the engine to detonate. Coolant temperature sensors are called thermistors, which are used to measure temperature. There are two types of thermistors: the positive temperature coefficient (PTC) and the negative temperature coefficient (NTC). Both types are resistors that change value when exposed to different temperatures. All conductors change resistance when their temperature changes and thermistors have been designed to maximize this change.

The PTC thermistor behaves like most conductors; as the temperature increases, its resistance increases. The NTC thermistor decreases resistance as the temperature increases. The thermistors used in Chrysler fuel injection and electronic timing control systems are NTC. The resistance specifications for the Chrysler coolant temperature sensor follow:

-40deg F   =   100,700 ohms
0deg F   =   25,000 ohms
20deg F   =   13,500 ohms
40deg F   =   7,500 ohms
70deg F   =   3,400 ohms
100deg F   =   1,600 ohms
160deg F   =   450 ohms
212deg F   =   185 ohms

The Coolant Sensor Circuit

The thermistor circuit behaves a little differently from the electrical circuits with which most automotive technicians are familiar. A power supply, 5 volts,

Older applications used a temperature controlled vacuum switch to limit or even eliminate the effects of the vacuum advance when the engine was cold. The EFI Chryslers use a coolant temperature sensor feeding information to the Logic Module, or SMEC. Also, when the ignition base timing is tested, the coolant temperature sensor must be disconnected.

supplies a reference voltage to the circuit. Before leaving the computer, the 5 volt current passes through a fixed value resistor causing a voltage drop. The current then continues through the thermistor and on to ground where the voltage is zero. As the resistance of the thermistor changes, the voltage on the wire between the fixed value resistor and the thermistor will also vary. The computer measures this voltage on the outbound side of the fixed value resistor to determine the temperature.

When the computer sees a comparatively high voltage on the wire to the coolant sensor, it knows that the resistance in the thermistor is high; therefore, the temperature of the sensor is low. A low voltage on this wire means that the resistance is low and therefore the temperature must be high.

Setting the Timing

To set the timing on a Chrysler fuel injected application, start the engine, allow it to come to operating temperature, and disconnect the coolant temperature sensor. The computer senses a problem in the coolant sensor input circuit and locks the timing at initial. After adjusting the timing, reconnect the coolant sensor, shut the engine off, and restart. The computer should now be back in the timing control mode.

Troubleshooting

For the rare reader who is curled up in front of a cozy fire reading this manuscript cover to cover, this is the first system mentioned where using the onboard diagnostic abilities of the computer in troubleshooting ignition problems makes sense.

No Start: Basic Tests

During the past six years, I have trained hundreds of technicians in the fine art of no-starts. The fuel injected Chrysler applications are both the easiest and hardest no-starts to troubleshoot. The following procedure is virtually foolproof, however.

Remove a spark plug wire from any of the spark plugs, insert a screwdriver with a plastic handle in the end of the plug wire so that the metal shaft is a quarter inch from a good ground. Have someone crank the engine and look for a spark. If there is a spark, the no-start problem is not related to the ignition system. Check the condition of the timing belts or chains, compression, and spark plugs. If there is no spark, remove the coil wire from the distributor cap and insert the screwdriver as in the plug wire. Crank the engine. If there is a spark, replace the distributor cap and rotor. If there is no spark, turn off the ignition switch, reconnect all the spark

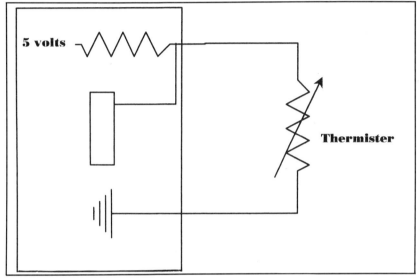

The coolant temperature sensor is a negative temperature coefficient (NTC) thermistor. As the temperature in the sensor is exposed to increases, the voltage on the 5 volt reference wire drops. A high voltage, say 4.2 volts, would indicate that the coolant temperature is low; when the engine reaches operating temperature, it will be much lower, about 1.2 volts.

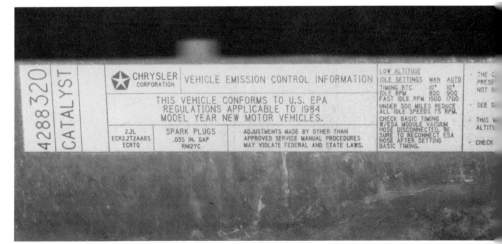

The only dependable way to find the correct ignition timing specification is to read the EPA label located under the hood.

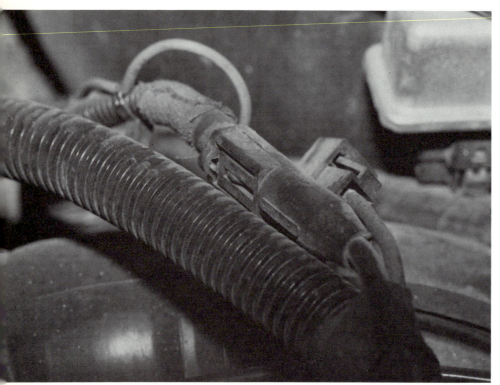

When trouble codes have been set in the Logic Module, or SMEC, they can be erased by locating and disconnecting this connector. It is also the connector you disconnect to begin the diagnostic service procedure for no-starts when there is no spark from the ignition coil.

The three terminals connect the Hall Effects to the vehicle wiring harness. One of these pins should receive power from the Power Module, or SMEC, another should provide a ground, and the last should deliver a square wave from the Hall Effects to the Logic Module, or SMEC.

plug wires and coil wire, then disconnect the battery.

After the battery has been disconnected for about ten seconds, reconnect it and crank the engine for ten seconds. During the time the engine is being cranked, the computer is seeking a signal from the Hall Effects or optical sensor. If the computer fails to see this signal, the computer will generate a Code 11. When the engine has cranked for ten seconds, turn the key to the off position. Pause about five seconds and rotate the key on and off three times, ending in the on position. Near the center of the instrument, a small red or amber trouble light reading either "Power Loss" or "Check Engine" will begin to flash. Count the flashes, they will be arranged in a two-digit format. Specifically, you are looking for a Code 11. This will be seen as one flash followed by a short pause and then another single flash.

Observe and record all codes; they may be helpful in diagnosing. Other codes that relate to no-start conditions are Code 13 for the vacuum supply to the MAP sensor; Code 14 for an electrical problem with the MAP sensor; Code 22 for the coolant sensor (especially if the engine starts okay when the engine is cold); and Code 42 for the auto-shutdown (ASD) relay.

If there is a Code 11, replace the distributor pickup. If this does not fix the problem, or if there is no Code 11, inspect the wiring from the distributor to the computer. Refer to the appropriate procedure below:

1984 Throttle Body and Multipoint Injected

Disconnect the three pin connector at the distributor. Turn the ignition switch on and check the voltage in the orange wire. If there are less than 6 volts, check continuity through the orange wire back to the number 12 terminal of the 12 pin connector on the power mod-

The SMEC has a 60-pin connector and a 14-pin connector. The 14-pin connector handles the inputs and outputs that the Power Module handles in the two module systems. The 60-pin connector handles those that were connected to the Logic Module.

The Chrysler onboard computer system can communicate with a diagnostic tool called a Diagnostic Readout Box (DRB) or DRB2. This tool plugs into a connector located under the hood. Among the things that the computer can communicate to the DRB are trouble codes, coolant temperature, manifold pressure, and engine speed.

ule. If the wire is in good condition, confirm there is 12 volts to the connection of the dark blue wire at terminal 2 of the Power Module 10 pin connector. If there are 12 volts to terminal 2, replace the Power Module. Once there are close to 8 volts on the orange wire at the distributor, check for a good ground on the black/light blue wire at the distributor connector. If this wire has a good ground to the engine block and/or chassis, check continuity between the terminal, distributor harness end of the gray wire, and pin 7 of the 12 pin connector of the Power Module. Also check for continuity from pin 7 of the 12 pin connector of the Power Module and terminal of the light colored connector on the Logic Module. Once the condition of the wires has been confirmed, the problem is either the power module or the distributor pickup.

With the distributor connector disconnected but all the computer terminals connected, place the distributor cap end of the coil wire a quarter inch from ground and turn the ignition switch to the on position. Momentarily, place a jumper wire between the gray and light blue wires on the vehicle harness for the distributor. As contact is made and broken, the coil wire should arc to ground. If the coil wire does arc, replace the Hall Effects pickup in the distributor. If the coil wire does not arc to ground, replace the Power Module.

1985-1987 Throttle Body and Multipoint Injected

Disconnect the three pin connector at the distributor. Turn the ignition switch on and check the voltage in the orange wire. If there are less than 6 volts, check continuity through the orange wire back to the number 12 terminal of the 12 pin connector on the Power Module. If the wire is in good condition, confirm there are 12 volts to the connection of the dark blue wire at terminal 2 of the Power Module 10 pin connector. If there are 12 volts to terminal 2, replace the Power Module. Once there are close to 8 volts on the orange wire at the distributor, check for a good ground on the black/light blue wire at the distributor connector. If this wire has a good ground to the engine block and/or chassis, check continuity between the terminal, distributor harness end of the gray wire, and pin 10 of the light colored connector on the Power Module. Once the condition of the wires has been confirmed, the problem is either the Logic Module, the Power Module, or the distributor pickup.

With the distributor connector disconnected but all the computer terminals connected, place the distributor cap end of the coil wire a quarter inch from ground and turn the ignition switch to the on position. Momentarily, place a jumper wire between the gray and light blue wires on the vehicle harness for the distributor. As contact is made and broken, the coil wire should arc to ground. If the coil wire does arc, replace the Hall Effects pickup in the distributor. If the coil wire does not arc to ground, check continuity through the yellow wire that runs from terminal 10 of the Power Module's 12 pin connector. If the wire is good, connect an analog voltmeter to the yellow wire connected to terminal 10 of the 12 pin connector on the Power Module. Crank the engine and look for fluctuations on the needle. If the needle fluctuates, replace the Power Module. If the needle does not fluctuate, replace the Logic Module.

1987-1990 SMEC System, 3.0 Liter Optical and All Hall Effects

Disconnect the three pin connector at the distributor. Turn the ignition switch on and check the voltage in the orange wire. If there are less than 6 volts, check continuity through the orange wire back to the number 1 pin of the SMEC 14 pin connector. If the wire is in good condition, confirm there are 12 volts to terminals 12 and 41 of the SMEC 60 pin connector and pin 4 of the 14 pin connector. If there are 12 volts on each of these terminals, replace the SMEC. Once there are close to 8 volts on the orange wire at the distributor, check for a good ground on the black/light blue wire at the distributor connector. If this wire has a good ground to the engine block and/or chassis, check continuity between the distributor harness end of the gray wire and pin 47 of the 60 pin connector. Once the condition of the wires has been confirmed, the problem is either the SMEC or the distributor pickup.

With the distributor connector disconnected but all the computer terminals connected, place the distributor cap end of the coil wire a quarter inch from ground and turn the ignition switch to the on position. Momentarily, place a jumper wire between the gray and the light blue wire on the vehicle harness for the distributor. As contact is made and broken, the coil wire should arc to ground. If the coil wire does arc, replace the optical pickup in the distributor. If the coil wire does not arc to ground, check continuity through the yellow wire that runs from terminal 13 of the 14 pin connector to terminal 34 of the 60 pin connector. If the wire is good, replace the SMEC.

Testing the Distributor Pickup

Here's a trick of the trade: The distributor pickup on modern electronic fuel injected cars does double duty as the reference pulse signal for the ignition system and for the triggering of the injector. If, in a no-start condition on a fuel injected car, there is no spark from the coil

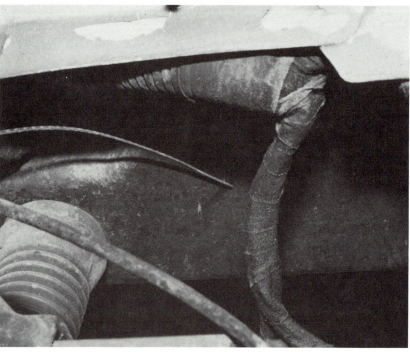

The SMEC is always located in the engine compartment. Sometimes it is well hidden.

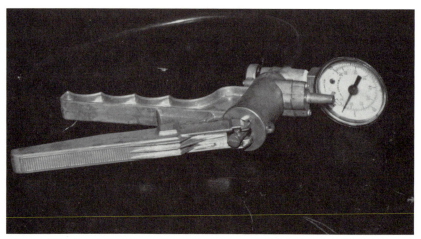

Connect a voltmeter to the center wire of the MAP sensor with the wiring harness still connected to the MAP sensor. Connect a hand-held vacuum pump. The vacuum reading should drop when vacuum is applied.

wire and the injector is not opening, the problem is most likely the distributor pickup.

Using Trouble Codes

Chrysler, like other domestic manufacturers, has an onboard diagnostic program built into the computer. For the aforementioned test to determine if the pickup Hall Effects is defective and responsible for the no-start condition, in addition to, or instead of, the Code 11, several other codes may have been obtained. Codes that may relate to the ignition system and the no-start condition are 13, 14, and 42.

Code 13

A Code 13 without a Code 14 means that there is a restriction in the vacuum supply line from the intake manifold to the MAP sensor. Although the MAP sensor has an effect on the ignition timing, it has a much greater effect on the fuel injection system. Besides the poor power problem, a restriction in the vacuum hose to the MAP sensor will cause the injection system to run extremely rich.

Code 13, 14

When these codes appear at the same time, the MAP sensor circuit has suffered an electrical failure. The MAP sensor may be defective or it may be the MAP sensor wiring harness or connector. While this sensor does affect ignition time, it is no more likely to cause a no-start than is a defective vacuum advance on a non-computer controlled ignition system.

Code 42

The presence of a Code 42 may mean that there has been intermittent problems with the Hall Effects in the past. Code 42 can also result from a near-stall condition. With regards to the no-start condition, a Code 42 could mean a problem with the ASD relay circuit.

To troubleshoot the Code 42 as a possible cause of the no-start condition, begin by locating the ASD relay. It is located above the Logic Module on 1984 models, next to the Power Module on 1985 and 1986 models, in the Power Module during the 1987 model year, and next to the SMEC on the 1988 and later models. The ASD relay can be distinguished from other relays by the wiring colors.

There are four terminals on the ASD relay that have wires leading to them. One wire is red; it runs to the battery through a fusible link. Another is dark green/black and is the output wire of the ASD relay. The dark green/black wire carries battery voltage to the injectors, coil positive, the canister purge solenoid,

The black relay is the auto-shutdown (ASD) relay.

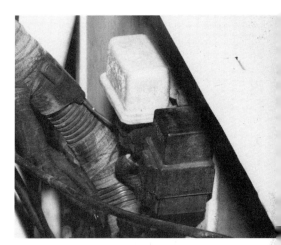

One dealership the author worked at recommended replacing the Hall Effects at 60,000 miles. It seemed that the failure rate for this sensor increased dramatically after this mileage. Failure is usually first noticed as a tendency for the engine to die intermittently.

the air switching solenoid, EGR solenoid, the alternator field, and the fuel pump. As you can see, when the ASD relay fails to operate properly, almost everything else does, also. Check for a voltage on the ASD terminal of the dark green/black wires while the engine is being cranked. If there is a voltage, check for power to the injectors, coil positive, the canister purge solenoid, the air switching solenoid, EGR solenoid, the alternator field, and the fuel pump.

If all of these have power, the Code 42 is not related to the no-start problem. If any do not have power, repair the wires as necessary. If there is no power on the dark green/black wire, proceed with this procedure. A dark blue wire carries switched ignition voltage to the pulldown windings of the ASD relay. The ASD relay is activated when the computer grounds the pull down winding of the ASD though the blue/yellow wire.

Begin with the ignition switch off. Check for voltage at the red wire. There should be 12 volts. Turn the ignition switch on and check for voltage at the dark blue wire. There should be 12 volts. Repair these wires as necessary. There should be battery cranking voltage while the engine is being cranked at the blue and yellow wire. If there is not cranking voltage, check the continuity of the blue and yellow wire to terminal 58 on the SMEC. If the wire is good and there is no voltage at the ASD relay while the engine is being cranked, replace the SMEC. If there is voltage at the dark blue/yellow wire as it enters the ASD relay but there is no voltage on the dark green/black wire, replace the ASD relay.

Starts But Does Not Continue to Run When the Key Is Released

It is highly unlikely that this symptom could be caused by this ignition system.

Dies While Driving Down the Road

The most likely cause of this symptom is an intermittent open in the Hall Effects. As has been mentioned before, intermittent problems are often the hardest to find. Unfortunately, there is no way to confirm an intermittently defective component in this system. Before spending the money on a new Hall Effects, be sure to check all the wires and connectors that relate to the ignition system and fuel injection system. Other components that can cause the engine to die intermittently include the Logic Module, Power Module, or SMEC as well as the injector and ignition switch.

Misfire at Idle

This is almost always never the fault of the primary ignition system; it is almost always the fault of some component in the secondary side of the ignition system.

Before troubleshooting any misfire, it is essential to verify that the engine is in good condition. A compression test is a good starting point. If the valves are adjustable, be sure that they are properly adjusted.

With a pair of "sissy" pliers, remove and replace one plug wire at a time from the spark plugs. As each plug wire is removed, the engine rpm should drop. If one of the cylinders fails to produce as great a drop in rpm as the others, that cylinder is the source of the misfire.

Assuming the cylinder is in good condition and the valves are properly adjusted, remove the spark plug wire for that cylinder and check the resistance. The resistance should be less than 10,000 ohms per volt. If the resistance is correct, replace the spark plug. Unless the spark plugs are very new, replace them all.

Misfire Under a Load

Assuming the engine is in good condition, begin troubleshooting this problem by checking the spark plug gap. If they are gapped properly, replace the spark plugs. Even new spark plugs can misfire under a load.

If replacing the spark plugs does not solve the problem, remove the distributor cap. Inspect the wiring to the points. Frayed wiring can cause an intermittent open circuit as the vacuum advance moves the breaker plate. The intermittent open can cause a misfire.

Lack of Power

The ignition timing is the single part or function of the ignition system that can cause power problems. To check the base ignition timing on the fuel injected Chrysler applications, start the engine and allow it to warm up. Disconnect the coolant temperature sensor and check the timing. If the initial timing is correct, check for trouble codes. The computer uses the MAP and coolant sensors and the Hall Effects to advance and retard the ignition timing. Trouble codes 13, 14, and 42 relate to these components.

Ford Solid State Ignition (SSI) 12

In 1974, Ford Motor Company introduced the Solid State Ignition (SSI). It was the company's first serious attempt at electronic ignition. While the system was successful, many of the early applications had an annoying habit of shutting down unexpectedly. I once had a customer who drove from Amarillo, Texas, to Seattle, Washington, with a defective SSI module. In Denver, he started counting the number of times the ignition module left him sitting on the side of the road. By the time he got to Seattle, the car had shut off sixty times. To top off his frustration, replacing the ignition module fixed the problem and only cost him $45.00.

Primary Ignition Components
Electronic Control Unit

Six wires are connected to the ECU on all except the 1982 through 1987 products. The 1982-1987 models might have two additional wires connected to the engine control computer. Through these two wires, the Engine Control Module (ECM) controls the ignition timing. The ECU replaces the points and ballast resistor of the conventional point/condenser ignition system.

In the six wires there are two wire and four wire connectors. The two wire connector will have a white or light blue wire and be connected to either the ignition switch or the starter solenoid. When the ignition switch is in the crank position, the ECU receives voltage from this wire. This allows full voltage to module when the engine is being started. The second wire of the two wire connector will be red or red/white. This wire carries full

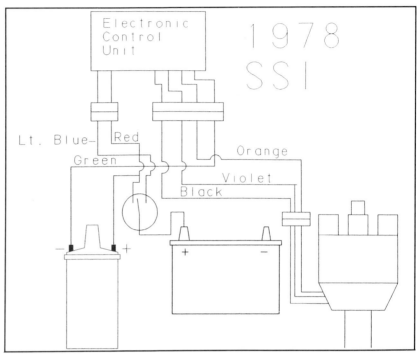

The Ford SSI system was introduced in 1974. Used by AMC as well as Ford, it had a thermal shutdown problem that caused intermittent stalling. Replacement of the ignition module cured the problem.

system voltage to the ECU while the engine is running. The four wire connector has a black, violet, orange, and dark green wire. The orange and violet wires are connected to a reluctance pickup in the distributor. The black wire is a ground while the dark green wire is connected to the negative terminal of the coil.

As the ECU receives pulses from the distributor pickup, it switches the ground for the ignition coil on and off. When the ground for the ignition coil is switched off, the magnetic field built by the current flow through the primary collapses. As the magnetic field collapses, it induces a high voltage in the secondary.

For the models that have the extra two wire connector, a yellow wire runs from the ECU to the ECM. This wire carries a signal from the ECM to the ECU to control the ignition timing. The extra wire out of the ECU goes only as far as the connector.

Distributor Pickup

The distributor pickup is a variable reluctance transducer. This sensor produces an alternating current as the distributor rotates. The sensor consists of a coil of wire, permanent magnet, and reluctor wheel. The reluctor wheel is part of the distributor shaft. As the reluctor wheel rotates, it distorts the magnetic field back and forth across the coil of wire. This distorting induces an alternating current in the coil of wire. As the speed of distributor shaft rotation increases, the amplitude (voltage output) of the sensor increases. More important, since the AC signal is a sine wave, the frequency of the sine wave changes as the speed of the distributor shaft changes. The frequency of the sine wave tells the ignition module the speed and position of the distributor shaft. Since the distributor shaft is synchronized to the cam shaft, the ignition is synchronized to the mechanical components of the engine. The ECU uses the signal from the pickup coil as a trigger source for the firing of the coil.

Ignition Switch

The ignition switch controls whether current will be available to the coil and ignition module. Unlike the point/condenser ignition system, the SSI system uses the same current path when starting the engine that it uses when the engine is running. There is no ballast resistor in the SSI system.

Secondary Ignition Components
Coil

The ignition coil secondary consists of hundreds of windings of very thin wire. When current stops flowing through the primary, the magnetic field created by the primary current flow collapses. This collapsing magnetic field induces several thousand volts in the secondary. This is the voltage that is used to jump the gap of the spark plugs to fire the mixture in the cylinders.

Because of the relatively low current in this high-voltage side of the coil, there are relatively few problems in the secondary side of the coil. The problems that do occur are usually in the form of open circuits.

Coil Wire

The coil secondary output wire carries the high-voltage current from the coil to the distributor cap. The coil wire is normally 6 to 12in long and has a resistance of a few thousand ohms. This relatively high resistance helps to reduce the intensity of the radio signal created by the secondary ignition.

All arcing generates a radio signal. In addition to the arcing that occurs at the spark plug, there is also an arc inside the distributor cap between the cap and the rotor. All of the secondary ignition wiring has a high resistance to reduce the affect of this arcing.

The coil wire can suffer from several possible problems. As the coil wire ages its resistance tends to increase. Also, as the wire ages the insulating quality of the jacket decreases, which makes it possible for the high voltages being carried by the wire to penetrate the jacket and arc to ground. Corrosion can also affect the current carrying ability of the wire.

A defective coil wire can result in misfiring, no-start, and poor power.

Distributor Cap and Rotor

These two components operate as a team. The coil wire delivers the high voltage to the center terminal of the distributor cap. A carbon conductor carries the voltage to the center of the rotor. The rotor will have either a metal or carbon resistive conductor that carries the voltage to the tip of the rotor. The rotor mounts on the top of the distributor shaft and is driven by the camshaft. As the rotor rotates, it approaches either copper or aluminum conductors on the inside of the distributor cap and arcs to these conductors, which carry the voltage to the spark plug wires.

The distributor cap is prone to cracking, corrosion, and carbon tracking. Carbon tracking occurs when a microscopic crack or piece of dirt provides a current path to ground that is easier than the current path and plug wire. The rotor is subject to corrosion and perforation. Perforation occurs when the high voltage seeks and finds a ground through the rotor to the distributor shaft.

Routine replacement of the distributor cap and rotor can prevent unforeseen problems. It is not necessary to replace the cap and rotor at each tune-up as many professional technicians recommend, but they should be replaced at every other tune-up. As you read this, however, do not assume that the preceding statement means that your mechanic has been ripping you off for the past 10 years. There is no disservice in a mechanic charging you $30 or $40 extra at each tune-up to ensure that you have less chance of developing premature problems.

As discussed in Chapter 2, when replacing the distributor cap or rotor, I advise that you replace them as a set. I further advise that they be built by the same name brand manufacturer. I have had situations in the past where a mismatched cap and rotor caused the rotor air gap to be so large that the engine either failed to start or misfired.

Spark Plug Wires

Back in the late sixties and the early seventies, when I first got into the car repair business, we checked spark plug wires by starting the engine in a darkened shop. If there were sparks flying around under the hood, it meant that one or more of the plug wires had perforated and was arcing to ground. For several years in the mid-seventies, I worked almost exclusively on fuel injected imports. Upon returning to working on domestics in 1980 I was mystified at the poor quality of the plug wires. I remembered few problems in the seventies, now there were many problems.

The difference was not the quality of the wires, but rather the leaner air/fuel ratios demanded in the early eighties (and today). In the early seventies, we unwittingly compensated for defective spark plug wires by enriching the idle mixture screws on the carburetors. This option was not available on the sealed carburetors of the late seventies. Therefore, technicians were forced to replace defective spark plug wires rather than simply masking the problem with a richer mixture.

When replacing spark plug wires, the old adage, "You get what you pay for" is especially true. A $50 set of spark plug wires can easily outlast four $12 sets of wires. A marginal plug wire can cause a misfire when the engine is under an extreme load and on a modern fuel injected car can make the engine run rich.

Spark Plugs

Every tune-up includes replacing the spark plugs. There are many brands of spark plugs on the market, some good, some bad. Asking for opinions on which is the best brand of spark plug is like asking a group which is the best soft drink, it is largely a matter of personal opinion. What I have always done, when I had a choice, was to use the brand that the manufacturer installed at the factory. My thinking is that the manufacturer has a vested interest in choosing the spark plug that would provide the best driveability and has the least chance of requiring replacement within 12,000 miles. This method has rarely failed to provide either myself or my customers with good service.

The spark plug consists of a pair of electrodes separated by an air gap of between 0.028 and 0.075in. As the spark from the coil travels down the plug wire seeking ground, it must arc across this air gap. If this arc is exposed to a properly preheated, well atomized mixture of fuel and air, this spark will ignite the fuel.

As the spark plugs arc at a high frequency in high temperatures, the electrode material will slowly vaporize. This causes the gap the widen. The wider the gap, the higher the voltage required to initiate the spark across the gap. Eventually, the voltage required to initiate the spark across the gap will be greater than the coil is capable of generating and a misfire will occur.

Common problems associated with the spark plugs are misfiring and difficulty in starting.

Timing Control

The ignition timing must change as the engine is running to adjust to different engine speeds and loads. When the engine is running at an idle, the spark must begin at a point in crankshaft rotation that will allow for the spark to extinguish when the crankshaft is about 10deg after top dead center. Since the length of time the spark is jumping the gap is a relative constant, about 2.5 milliseconds, the spark must start sooner as the engine speed increases.

Centrifugal Advance

As an example, on a hypothetical engine the spark occurs at 10deg before top dead center (TDC) when the engine is running at 1000rpm. At this speed, the spark extinguishes at about 10deg after TDC. This means that the crankshaft has rotated 20deg since the initiation of spark. As the engine speed increases, the crankshaft rotates more degrees in the 2.5 milliseconds that the spark is jumping the gap. At 2000rpm the crankshaft will rotate twice as much. If the timing at 1000rpm should be 10deg before TDC, then the timing at 2000rpm should be about 30deg before TDC. As the engine speed continues to increase, the timing will need to continue to advance. The amount of total advance, the upper limit of the advance, will vary depending on the design of the engine.

The change of timing in response to rpm is accomplished through a set of spring-loaded weights. As the speed of the engine increases, the weights swing out against spring tension. The trigger wheel that trips the sensor, although it is mounted on the distributor shaft, is not part of the distributor shaft. The swinging weights cause the trigger wheel to rotate with respect to the distributor shaft. This advances the timing.

Vacuum Advance

At first glance this is a misnomer. The vacuum advance actually retards the timing when the engine is under a load. In most applications the vacuum advance is connected to ported vacuum. The advance unit receives no vacuum at an idle, but when the throttle is opened the vacuum advances the timing. As the load on the engine increases, the vacuum drops. As the vacuum drops, the timing is not advanced as much, it retards. Retarding the timing lowers the combustion temperature, therefore prevents detonation and decreases the potential of damage to the engine.

Setting the Timing

When setting the ignition timing, it is best to follow the instructions outlined on the EPA sticker under the hood. If the instructions are missing or obliterated, disconnect the vacuum advance and adjust to the specification listed in this book.

Troubleshooting
No Start

Like most ignition systems, a no-start condition in the SSI system is among the easiest problems to troubleshoot. There are three things required to get the engine started: air, fuel, and spark. Since the topic of this book is ignition systems, I will concentrate on the ignition prob-

If the reluctor wheel becomes damaged, it is easily replaced. Note the timing mark on the top tooth.

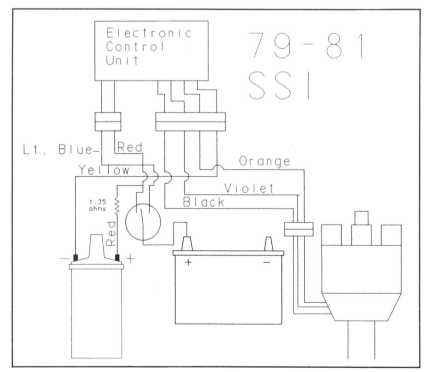

In 1979, a ballast resistor was added to the primary side of the coil circuit. This resistor reduced the amount of heat generated by the system.

lems that can prevent the engine from starting.

Begin testing the ignition system by disconnecting a plug wire from one of the spark plugs. Stick a plastic handled screwdriver in the end of one of the plug wires. While holding the screwdriver near ground, have someone crank the engine. If there is no spark, remove the coil wire from the distributor cap and place the screwdriver in the end of the coil wire. Hold the screwdriver a quarter inch from ground. Again, have someone crank the engine. If there is a spark, replace the distributor rotor and cap.

If there is no spark, connect a test light to the negative terminal of the coil. While you watch the test light, have someone crank the engine. If the test light blinks, the secondary windings of the ignition coil are defective and the coil should be replaced. To verify that the secondary windings are defective, use an ohmmeter to check the resistance of the secondary. The resistance at 75deg F should be between 7,700 and 9,300 ohms. If it does not blink there is a problem in the primary side of the ignition system.

Checking the Pickup Coil

Begin troubleshooting the primary side of the ignition system by checking the pickup coil. There are several viable ways to test the pickup coil, some are more thorough than others. One of the most commonly suggested methods is to use an ohmmeter. The resistance between the orange and violet wire through the pickup coil should be between 400 and 800 ohms. If the resistance through the pickup coil is considerably different from these readings, replace it. If the resistance is less than 100 ohms outside of these specifications, the no-start problem is likely to have another cause.

The ohmmeter method of testing the pickup coil is incomplete. While it does test the condition of the coil of wire, which is the most likely part of the pickup coil to be defective, it does not test the condition of the permanent magnet or the air gap between the pickup and the reluctor wheel. A better test is to disconnect the connector near the distributor or ignition module that contains the orange and violet wires. Connect an AC voltmeter between these wires on the distributor side of the connector. Have someone crank the engine while you watch the voltmeter. If the pickup generates less than 0.5 volt, replace the pickup. In other ignition systems, it would be important to check the air gap between the pickup and the reluctor wheel. However, in the SSI system this air gap is nonadjustable. If the air gap is incorrect, it would require replacing the pickup coil.

Testing the Module

There is no convenient way to test the ignition module directly, therefore it is best tested through process of elimination. If a test light connected to the negative side of the coil does not blink when the engine is cranked, and if the pickup tests good, replace the module.

Starts But Does Not Continue to Run When the Key Is Released

Like a point/condenser ignition system, the SSI system uses a ballast resistor wire to limit current flow through the primary when the engine is running. A bypass circuit allows full current flow when the engine is being cranked.

Connect a test light to the positive terminal of the coil and turn the ignition switch to the run position. If the light is not on, crank the engine. If the light is on while the engine is being cranked but goes out when the key is released, test the ballast resistor wire. The ballast resistor wire should have between 1.3

The SSI system uses an AC pickup coil located in the distributor. This pickup coil should produce 1/4 volt or more when the engine is cranking.

and 1.4 ohms of resistance. If the resistance is outside of this specification, replace the ballast resistor.

If the ballast resistor wire is good, or if replacing the ballast resistor did not solve the problem, inspect the wiring from the crank position of the ignition switch to the ignition terminal of the starter solenoid. If the ballast wiring is good, or if the repairs do not cure the symptom, replace the ignition module.

Dies While Driving Down the Road

In my experience with this system, an intermittent shutdown is one of the more common problems. By observation, the severity of the problem and the frequency of the problem seemed to be heat related.

Although I caution the reader that the pickup coil could be at fault and suggest that an AC voltage output test be done, the symptom is almost always caused by the module. What I suggest to professional technicians is that they order the ignition module, get it on its way, then test the pickup coil. If the pickup coil is found to be defective, most parts houses will accept an unopened box with an electrical or electronic component.

Misfire at Idle

This is almost always never the fault of the SSI system; it is almost always the fault of some component in the secondary side of the ignition system.

Before troubleshooting any misfire, it is essential to verify that the engine is in good condition. A compression test is a good starting point. If the valves are adjustable, be sure that they are properly adjusted.

With a pair of "sissy" pliers, remove and replace one plug wire at a time from the spark plugs. As each plug wire is removed the engine rpm should drop. If one of the cylinders fails to produce as great a drop in rpm as the others, that cylinder is the source of the misfire.

Assuming the cylinder is in good condition and the valves are properly adjusted, remove the spark plug wire for that cylinder and check the resistance. The resistance should be less than 10,000 ohms per volt. If the resistance is correct, replace the spark plug. Unless the spark plugs are very new, replace them all.

Misfire Under a Load

Assuming the engine is in good condition, begin troubleshooting this problem by checking the spark plug gap. If they are gapped properly, replace the spark plugs. Even new spark plugs can misfire under a load.

If replacing the spark plugs does not solve the problem, remove the distributor cap. Inspect the wiring to the points. Frayed wiring can cause an intermittent open circuit as the vacuum advance moves the breaker plate. The intermittent open can cause a misfire.

Lack of Power

There are many things that can cause a lack of power, some related to the ignition system, some not. Begin checking this

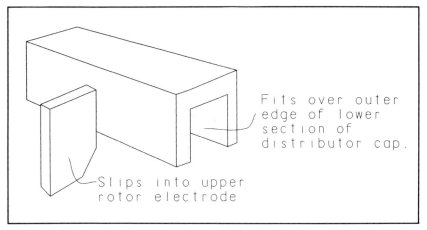

The Duraspark III ignition system was unique to Ford applications with the EEC-II and EEC-III engine control system.

problem by confirming the engine is in good condition as are the air and fuel filters.

If a lack of power is the result of problems in the ignition system, it is likely the problem is in the timing control system. To test the timing control system, connect a timing light to the engine. Disconnect the vacuum advance and plug in the hose. With the engine at idle speed, check the timing. Now raise the engine speed to 2000 to 2500rpm. If the timing does not advance, the centrifugal advance system is not working. Inspect the distributor weights. If they are free and move easily, replace the weight springs. If the springs are weak, they will allow the timing to advance all the way prematurely, even at idle. If the weights are frozen, use penetrating oil or whatever is necessary to free them. If they are badly corroded, it may be necessary to replace the distributor.

If, or when, the centrifugal advance is working properly, with the engine still at 2000 to 2500rpm, reconnect the vacuum hose to the vacuum advance. When the vacuum hose is reconnected, the timing should advance several degrees.

Duraspark III

13

During the late seventies and on into the eighties and nineties, a wide range of applications used the Ford Duraspark III ignition system. The system was used on passenger cars from 1979 through 1984; in 1982 and 1983 in the light-duty trucks. This ignition system is only used with the EEC-II and EEC-III engine control system.

Primary Ignition Components
Electronic Control Unit
The electronic control unit is a metal box located on the inner fender.

Pickup Coil
The pickup coil is a standard AC pickup. However, the location and how it receives the signal are unique. The Duraspark III system uses the tach reference signal from the on-board computer. The pickup coil is located either on the harmonic balancer or on the flywheel, depending on the engine. The AC signal is then sent to the ECU where the AC signal is converted to a square wave. The ECU uses this as a tach signal and passes the signal along the orange wire to the ignition module.

To see if the module is receiving the signal from the computer, connect a digital tachometer to the orange wire and crank the engine. This location means that there are no primary ignition components inside the distributor.

Ignition Switch
The ignition switch controls whether current will be available to the coil and ignition module.

Secondary Ignition Components
Coil
The ignition coil secondary consists of hundreds of windings of very thin wire. When current stops flowing through the primary the magnetic field created by the primary current flow collapses. This collapsing magnetic field induces several thousand volts in the secondary. This is the voltage that is used to jump the gap of the spark plugs to fire the mixture in the cylinders.

Because of the relatively low current in this high-voltage side of the coil, there are relatively few problems in the secondary side of the coil. The problems that do occur are usually in the form of open circuits.

Coil Wire
The coil secondary output wire carries the high-voltage current from the coil to the distributor cap. The coil wire is normally 6 to 12in long and has a resistance of a few thousand ohms. This relatively high resistance helps to reduce the intensity of the radio signal created by the secondary ignition.

All arcing generates a radio signal. In addition to the arcing that occurs at the spark plug, there is also an arc inside the distributor cap between the cap and the rotor. All of the secondary ignition wiring has a high resistance to reduce the affect of this arcing.

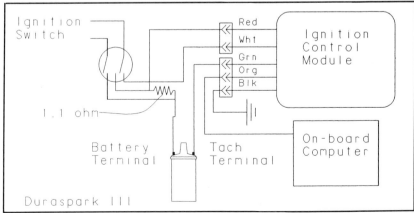

A unique characteristic of the SSI system is the need to "time" the distributor rotor. A special tool can be obtained from the dealer for this purpose.

The Duraspark ignition module is typically located on an inner fender well.

The coil wire can suffer from several possible problems. As the coil wire ages its resistance tends to increase. Also, as the wire ages the insulating quality of the jacket decreases, which makes it possible for the high voltages being carried by the wire to penetrate the jacket and arc to ground. Corrosion can also affect the current carrying ability of the wire.

A defective coil wire can result in misfiring, no-start, and poor power.

Distributor Cap and Rotor

These two components operate as a team. The coil wire delivers the high voltage to the center terminal of the distributor cap. A carbon conductor carries the voltage to the center of the rotor. The rotor will have either a metal or carbon resistive conductor that carries the voltage to the tip of the rotor. The rotor mounts on the top of the distributor shaft and is driven by the camshaft. As the rotor rotates, it approaches either copper or aluminum conductors on the inside of the distributor cap and arcs to these conductors, which carry the voltage to the spark plug wires.

The distributor cap is prone to cracking, corrosion, and carbon tracking. Carbon tracking occurs when a microscopic crack or piece of dirt provides a current path to ground that is easier than the current path and plug wire. The rotor is subject to corrosion and perforation. Perforation occurs when the high voltage seeks and finds a ground through the rotor to the distributor shaft.

Routine replacement of the distributor cap and rotor can prevent unforeseen problems. It is not necessary to replace the cap and rotor at each tune-up as many professional technicians recommend, but they should be replaced at every other tune-up. As you read this, however, do not assume that the preceding statement means that your mechanic has been ripping you off for the past 10 years. There is no disservice in a mechanic charging you $30 or $40 extra at each tune-up to ensure that you have less chance of developing premature problems.

As discussed in Chapter 2, when replacing the distributor cap or rotor, I advise that you replace them as a set. I further advise that they be built by the same name brand manufacturer. I have had situations in the past where a mismatched cap and rotor caused the rotor air gap to be so large that the engine either failed to start or misfired.

Spark Plug Wires

Back in the late sixties and the early seventies, when I first got into the car repair business, we checked spark plug wires by starting the engine in a darkened shop. If there were sparks flying around under the hood, it meant that one or more of the plug wires had perforated and was arcing to ground. For several years in the mid-seventies, I worked almost exclusively on fuel injected imports. Upon returning to working on domestics in 1980 I was mystified at the poor quality of the plug wires. I remembered few problems in the seventies, now there were many problems.

The difference was not the quality of the wires, but rather the leaner air/fuel ratios demanded in the early eighties (and today). In the early seventies, we unwittingly compensated for defective spark plug wires by enriching the idle mixture screws on the carburetors. This option was not available on the sealed carburetors of the late seventies. Therefore, technicians were forced to replace defective spark plug wires rather than simply masking the problem with a richer mixture.

The Duraspark distributor cap is in two pieces. The upper part, which contains the electrodes, can be replaced without replacing the entire cap.

When replacing spark plug wires, the old adage, "You get what you pay for" is especially true. A $50 set of spark plug wires can easily outlast four $12 sets of wires. A marginal plug wire can cause a misfire when the engine is under an extreme load and on a modern fuel injected car can make the engine run rich.

Spark Plugs

Every tune-up includes replacing the spark plugs. There are many brands of spark plugs on the market, some good, some bad. Asking for opinions on which is the best brand of spark plug is like asking a group which is the best soft drink, it is largely a matter of personal opinion. What I have always done, when I had a choice, was to use the brand that the manufacturer installed at the factory. My thinking is that the manufacturer has a vested interest in choosing the spark plug that would provide the best driveability and has the least chance of requiring replacement within 12,000 miles. This method has rarely failed to provide either myself or my customers with good service.

The spark plug consists of a pair of electrodes separated by an air gap of between 0.028 and 0.075in. As the spark from the coil travels down the plug wire seeking ground, it must arc across this air gap. If this arc is exposed to a properly preheated, well atomized mixture of fuel and air, this spark will ignite the fuel.

As the spark plugs arc at a high frequency in high temperatures, the electrode material will slowly vaporize. This causes the gap to widen. The wider the gap, the higher the voltage required to initiate the spark across the gap. Eventually, the voltage required to initiate the spark across the gap will be greater than the coil is capable of generating and a misfire will occur.

Common problems associated with the spark plugs are misfiring and difficulty in starting.

Timing Control

All ignition timing decisions are made by the EEC-II or EEC-III computer.

Setting the Timing

Timing for the Duraspark III system is completely nonadjustable. However, a unique feature of this system is the need to adjust the rotor. While this does not significantly affect ignition timing, it has a significant effect on the resistance in the secondary ignition system and therefore a significant effect on the strength of the spark available to the combustion chamber.

The rotor design is also unique. Instead of having a single pickup in the center of the rotor, there are two pickups to route the secondary energy from the coil wire to two secondary distribution electrodes at opposite ends of the rotor.

Troubleshooting
No Start: Basic Tests

Use a tachometer to test for a pulse on the orange wire to the ignition module. If there is no pulse, replace the crank pickup. If there is still no pulse on the orange wire, replace the on-board computer.

If there is a pulse on the orange wire, check for 12 volts on the white wire while the engine is being cranked. Check for continuity to ground on the black wire. Check for 12 volts on the red wire with the key on and the engine off. If these voltages are not as described, repair the wires as necessary. If the wires are in good condition and the voltages are as described, connect a test light to the negative terminal of the coil. Crank the engine and check for a pulse on the test light. If there is no pulse, replace the ignition module. If there is a pulse, check for 12 volts on the positive terminal of the ignition coil. If there are 12 volts on the positive terminal of the coil, replace the ignition coil.

Testing the Crank Pickup: With an Ohmmeter

Most books on troubleshooting electronic ignition systems will suggest using an ohmmeter to test a reluctance pickup. Although this is not a totally invalid test, it only tests the coil of wire. Reluctor air gap, condition of the permanent magnet, and adequate rotational speed are not tested with this method.

Disconnect the pickup from the ignition module or vehicle wiring harness leads. Connect the ohmmeter to the pickup coil leads and measure the resistance. Typical ohmmeter readings for a good reluctance pickup coil would be between 500 and 1,500 ohms.

With an Oscilloscope

A much better way to test a reluctance pickup is with an oscilloscope. Connect the scope to the pickup coil leads and rotate the reluctor (crank the engine, rotate the wheel). A series of ripples should appear on the scope. If the line remains flat, there is an open in the coil of wire, the permanent magnet has been damaged, or the air gap between the pickup and the reluctor teeth is too large.

With an AC Voltmeter

The AC voltmeter is a practical and effective alternative to the oscilloscope. Connect the AC voltmeter in the same manner as described for connecting the oscilloscope. Rotating the reluctor at minimum speed (cranking the engine, rotating a wheel at one revolution per second) would yield between 0.5 and 1.5 volts. If the AC voltmeter does not produce a voltage, there is an open in the coil of wire, the permanent magnet has been damaged, or the air gap between the pickup and the reluctor teeth is too large.

Notes on Testing

Since the VRT is primarily a coil of wire and a permanent magnet, it is prone to intermittent failures with changes in temperature and vibration. If the failure being diagnosed is intermittent, the sensor should be heated and tapped while testing.

Starts But Does Not Continue to Run When the Key Is Released

Confirm that there are 12 volts on the white wire while the engine is being cranked. If there are not, repair the wire. If there are, replace the ignition module.

Dies While Driving Down the Road

Confirm that there are no loose or defective wires, and replace the ignition module.

Misfire at Idle

This is almost always never the fault of the Duraspark III system; it is almost always the fault of some component in the secondary side of the ignition system.

Before troubleshooting any misfire, it is essential to verify that the engine is in good condition. A compression test is a good starting point. If the valves are adjustable, be sure that they are properly adjusted.

With a pair of "sissy" pliers, remove and replace one plug wire at a time from the spark plugs. As each plug wire is removed, the engine rpm should drop. If one of the cylinders fails to produce as great a drop in rpm as the others, that cylinder is the source of the misfire.

Assuming the cylinder is in good condition and the valves are properly adjusted, remove the spark plug wire for that cylinder and check the resistance. The resistance should be less than 10,000 ohms per volt. If the resistance is correct, replace the spark plug. Unless the spark plugs are very new, replace them all.

Misfire Under a Load

Assuming the engine is in good condition, begin troubleshooting this problem by checking the spark plug gap. If they are gapped properly, replace the spark plugs. Even new spark plugs can misfire under a load.

If replacing the spark plugs does not solve the problem, remove the distributor cap. Inspect the wiring to the points. Frayed wiring can cause an intermittent open circuit as the vacuum advance moves the breaker plate. The intermittent open can cause a misfire.

Lack of Power

Since timing is controlled by the on-board computer, lack of power will not be caused by the Duraspark III system.

Thick Film Integrated (TFI) Ignition 14

When the EEC-IV fuel injection system was introduced in 1984, the Thick Film Integrated ignition system became Ford's standard ignition system. Two versions of the TFI system exist: the conventional and computerized timing controls; both are easy to identify. The TFI module on applications with conventional timing control have only three wires going into the end of the module. TFI applications with computerized timing control have six wires going into the end of the module.

Components

The PIP Sensor (Hall Effects)

The Hall Effects sensor is often used as an alternative to the pickup coil. Many ignition systems, both distributorless and distributor type, use a Hall Effects device. Its primary advantage over the pickup coil is its ability to detect position and rotational speed from zero rpm to tens of thousands. Its primary disadvantage is that it is not as rugged as the pickup coil and is more sensitive to errant magnetic fields. An intense magnetic field can shut down the proper operation of a Hall Effects.

How the PIP Sensor Works

The PIP sensor is a Hall Effects device. A Hall Effects device is a semiconductor that responds to the presence of a magnetic field. A current passes through the semiconductor from the positive lead to the negative lead. The semiconductor is positioned opposite a permanent magnet. A ferrous metal, windowed armature rotates between the permanent magnet and the semiconductor. When the win-

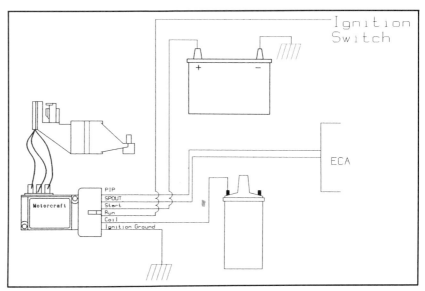

The Thick Film Integrated (TFI) ignition system is the only ignition system used on Ford fuel injected cars after 1983. The module and the fuel injection computer work together to control ignition timing.

A Hall Effects is used as the distributor pickup on the fuel-injected applications. Note that the ignition module is mounted on the distributor. On many late-model applications, the module is mounted on a fender well. Notice also the set of blades resting in front of the distributor, this is the rotary vane cup for the Hall Effects.

dow is open, the output voltage of the sensor is low. When the armature rotates and the window closes, the output voltage of the sensor drops low. The result of this is that as the crankshaft rotates, a square wave is generated, the frequency of which is directly proportional to the speed of crankshaft rotation.

Common Failures

The PIP sensor has an insulating connector block that connects the three terminals to the TFI module. In the early versions of the PIP sensor, this insulator block was rubber. Oil vapor would travel up the distributor shaft and deposit on the insulator block. These oil deposits would soften the insulator, making it conductive. Symptoms would include intermittent hesitation, stalling, and no start.

How the Hall Effects Works

A Hall Effects pickup is a semiconductor carrying a current flow. When a magnetic field falls perpendicular to the direction of that current flow, part of that current is redirected perpendicular to the main current path. The semiconductor is placed near a permanent magnet. A set of metal blades, or armature, attached to a rotating shaft or other device passes between the Hall Effects semiconductor and the permanent magnet. As the armature rotates, the magnet field is alternately applied to the Hall Effects and interrupted. The result is a pulsing current perpendicular to the main current path. This frequency is directly proportional to the speed of armature rotation. Since the output is only dependent on the presence of the magnetic field, the Hall Effects unit is capable of detecting armature position even when there is no rotational speed.

Testing

With an Ohmmeter

There is no valid test procedure on the Hall Effects using an ohmmeter.

With an Oscilloscope

Connect the oscilloscope to the Hall Effects signal lead. Rotate the armature. Depending on the number of blades and the rotational speed of the armature, the scope pattern could appear either as a square wave or a flat line that rises and falls with rotation.

With a Voltmeter

Connect a voltmeter to the Hall Effects output lead. The voltmeter should display either a digital high (4 volts or more) or a digital low (around 0 volt). Slowly rotate the armature while observing the voltmeter. If the voltmeter had read low, it should now read high; if the voltmeter had read high it should now read low. If the voltage fluctuates in this manner as the armature is rotated, then the Hall Effects is good.

With a Dwell Meter

Since the signal generated by the Hall Effects is a square wave, the dwell meter becomes a natural for testing. Connect the dwell meter between the Hall Effects output and ground. Rotate the armature as fast as possible (example, crank the engine); the dwell meter should read something besides zero and full scale. If it does, the Hall Effects is good.

With a Tachometer

As with the dwell meter, the tachometer is also a good tool for detecting square wave. Connect the tachometer between the Hall Effects output and ground. With

The Ford Hall Effects is called a PIP. In addition to being the pickup for the ignition system, it supplies the reference signal for the firing of the injector.

Fuel-injected models that open the injectors sequentially use a signature PIP to identify the number one cylinder firing position. The narrow blade on this rotary vane cup is the signature PIP.

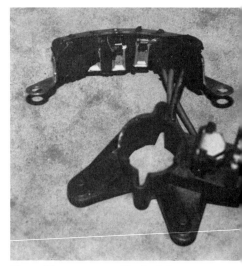

The rubber connector block of the early PIP sensors tended to deteriorate when it was exposed to the oil mist that traveled up the distributor shaft.

the armature rotating as described in the paragraph on the dwell meter, the tachometer should read something other than zero if the Hall Effects is good.

TFI Module

The TFI module does many jobs in this system. Like the points of a point/condenser ignition system, the TFI module is used to control current flow through the ignition coil. Additionally, it works with the Engine Control Assembly, or ECA (Ford's name for the fuel injection computer), to control the ignition timing.

Pickup Coil

A more scientific term for this device is variable reluctance transducer (VRT). The typical automotive technician will think of the electronic ignition distributor when this component is mentioned. In reality, there are many uses for this device. The pickup coil is used to measure crankshaft rotational speed on some distributorless ignition systems, wheel speed for four wheel anti-lock braking systems, differential speed on rear wheel antilock braking systems, and in some vehicle speed sensors. The pickup coil produces an AC sine wave with a frequency directly proportional to the speed of the rotation.

The primary advantage of the pickup coil over other rotational speed sensors is its simplicity. Consisting of a coil of wire, permanent magnet, and rotating reluctor, there is very little that can go wrong with it. Its main disadvantage is its inability to accurately detect low speed rotation. At low rotational speeds, the pickup coil is unable to produce a signal.

How the Pickup Coil Works

A coil of wire sits in a magnetic field created by a permanent magnet. A metal wheel with protruding reluctor teeth rotates through the magnetic field. As it rotates and one of the teeth approaches the magnetic field, the field is bent toward the approaching tooth. As it is bent, it passes across the coil of wire inducing a voltage. Continuing to rotate, the reluctor tooth drags the magnet field across the coil of wire eventually bending it in the opposite direction. The result is an AC signal.

Testing the Pickup Coil

With an Ohmmeter. Most books on troubleshooting electronic ignition systems will describe using an ohmmeter to test

The natural connector block on the right was introduced in 1986 and seemed to solve the oil saturation problem experienced by the black connectors.

Changing the PIP requires removing the distributor from the engine, then removing the rotary vane cup and distributor shaft from the distributor.

a reluctance pickup. Although this is not a totally invalid test, it only tests the coil of wire. Reluctor air gap, condition of the permanent magnet, and adequate rotational speed are not tested with this method.

Disconnect the pickup from the ignition module or vehicle wiring harness leads. Connect the ohmmeter to the pickup coil leads and measure the resistance. Typical ohmmeter readings for a good reluctance pickup coil would be between 500 and 1,500 ohms.

With an Oscilloscope

A much better way to test a reluctance pickup is with an oscilloscope. Connect the scope to the pickup coil leads and rotate the reluctor (crank the engine, rotate the wheel). A series of ripples should appear on the scope. If the line remains flat, there is an open in the coil of wire, and the permanent magnet has been damaged or the air gap between the pickup and the reluctor teeth is too large.

With an AC Voltmeter

The AC voltmeter is a practical and effective alternative to the oscilloscope. Connect the AC voltmeter in the same manner as described for connecting the oscilloscope. Rotating the reluctor at minimum speed (cranking the engine, rotating a wheel at one revolution per second) would yield between 0.5 and 1.5 volts. If the AC voltmeter does not produce a voltage, there is an open in the coil of wire, and the permanent magnet has been damaged, or the air gap between the pickup and the reluctor teeth is too large.

Notes on Testing

Since the pickup coil is primarily a coil of wire and a permanent magnet, it is prone to intermittent failures with changes in temperature and vibration. If

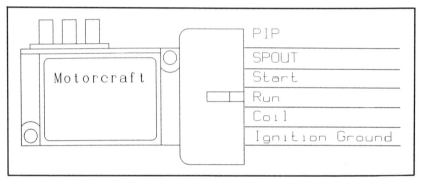

The top terminal is the PIP output to the Engine Control Assembly (ECA). The ECA uses this signal as an engine speed input and to synchronize the injectors. The second wire is the SPOUT (Spark Out) wire, which carries the timing control signal from the ECA back to the TFI module. The third wire provides power to the module when the starter is engaged; the fourth provides power when the engine is running. The fifth wire goes to the ignition coil to ground and fire the coil. The bottom wire is ground.

The ignition module that does not have the full metal backing was prone to failures when fluids leaked in around the edges of the metal.

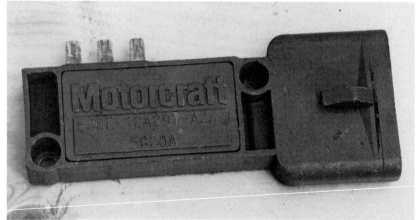

On applications where the TFI module mounts directly on the distributor, there are three terminals on the top of the module. These terminals connect directly to the PIP sensor.

The TFI ignition module on the bottom is used when the ignition module is mounted in a heat sink on an inner fender well or the rear engine compartment bulkhead.

the failure being diagnosed is intermittent, the sensor should be heated and tapped while testing.

E-Core Coil

The ignition coil secondary consists of hundreds of windings of very thin wire. When current stops flowing through the primary the magnetic field created by the primary current flow collapses. This collapsing magnetic field induces several thousand volts in the secondary. This is the voltage that is used to jump the gap of the spark plugs to fire the mixture in the cylinders.

Because of the relatively low current in this high-voltage side of the coil, there are relatively few problems in the secondary side of the coil. The problems that do occur are usually in the form of open circuits.

Prior to the mid-seventies, the ignition coil was a large oil-cooled component. Modern ignition coils are air cooled and referred to as an E-core coil.

SPOUT Wire and Connector

There is a yellow/green wire (referred to as Spark Out, or SPOUT) that the ECA uses to control the switching of the TFI (ignition) module to control timing. There should be a pulse on this wire. If the signal is present at the ignition module but no timing advance occurs, then the ignition module is defective.

If the SPOUT signal is not present, refer to the Code 18 diagnostics later in this chapter.

Profile Ignition Pickup (PIP)

The PIP sensor is located in the distributor on most engines, and located on the crankshaft on distributorless engines. There are three wires going to the PIP, one carries battery voltage to the sensor, another is the ground, and another sends the signal to the TFI (ignition) module. The PIP signal then splits and goes to the coil current control section of the TFI module and to the ECA. The ECA uses the PIP signal to synchronize the injectors. The signal is also used as the tach signal to the computer.

Troubleshooting
No Start

Remove the coil wire from the distributor cap. Stick a screwdriver in the end of the coil wire and hold a quarter inch from a good ground. Crank the engine. If there is a spark, replace the coil wire in the distributor cap. Remove a spark plug wire. Stick a screwdriver in the end of the plug wire and hold it a quarter inch from ground. Crank the engine. If there is a spark, the problem is either bad spark plugs or a problem with fuel sup-

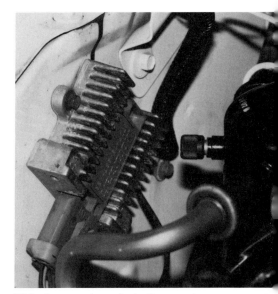

This Aerostar is typical of a remote-mounted TFI ignition module.

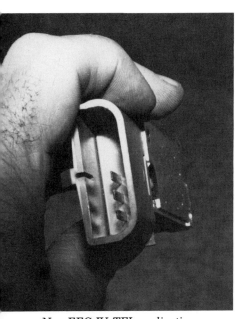

Non-EEC-IV TFI applications use a TFI module with only three terminals.

Non-EEC-IV TFI applications use an AC pickup instead of a Hall Effects.

109

ply.

If there is no spark at the end of the coil wire, connect a test light to the negative terminal of the coil. Crank the engine, does the test light blink? If not, the problem is either a bad TFI module or a defective PIP sensor. Connect a tachometer to the top wire of the TFI module (usually blue) and crank the engine. If the tachometer reads zero, confirm there is battery cranking volts to the third wire from the top while the engine is being cranked. Confirm the bottom wire has continuity to ground. If there is voltage to the third wire, and continuity to ground on the bottom wire, and no tach reading on the top wire, replace the PIP sensor. If there is a tach signal on the top wire, replace the ignition module.

There are times when a no-start problem and other driveability problems are best diagnosed with the help of the fuel injection computer. The EEC-IV fuel injection computer has the ability to detect circuit faults and lead the technician to the source of those faults by means of service codes. To get these codes, use the following procedure.

The EEC-IV STAR Test

To begin the EEC-IV test, locate the test terminal, which is in the engine compartment, under the hood. It is a six terminal connector usually found on one of the inner fender wells. Although there are pockets for six wires in this connector, there are normally only three or four of these wires in the connector. Located near the six pocket connector is a single pigtail connector. This is an essential part of the diagnostic hook-up.

Once the test connectors are located, the technician has a choice of several tools to connect to acquire the self-diagnostic information. Ford Motor Company provides its dealer service facilities with a dedicated tester known as the STAR tester. Several manufacturers of automotive tools and test equipment market testers to access the service codes. All of the specialized testers make the job of extracting the service codes easier and usually faster. However, these testers do not do anything that cannot be done with much simpler tools. My personal preference is to use an analog voltmeter to extract the codes. In my description of the use of the analog voltmeter, you could substitute the use of a test light or even a buzzer.

Ensure that the temperature of the engine is between 50 and 250deg F. If the temperature of

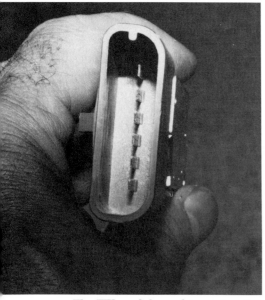

The TFI module used in conjunction with the EEC-IV engine control system has six terminals on the end connector.

When adjusting the ignition timing, the SPOUT connector should be removed from the wiring harness. This connector is typically located 6 to 8in from the ignition module. The wire color is either yellow/green or purple.

To obtain trouble codes from an EEC-IV TFI ignition system, locate the EEC-IV test connector. On late-model applications, the test connector is located inside a protective plastic cover.

the engine is outside this parameter, false codes related to engine temperature will be generated. Locate the test connector, connect the red lead of an analog voltmeter to the positive terminal of the battery. Connect the black lead of the voltmeter to the lower left of the center four terminals of the six pocket test connector. This terminal is known as the Self-test output (STO). At this point, the voltmeter should read 12 volts. Now connect a jumper wire between the upper right of the center four and the single pigtail wire. This single pigtail is known as the Self-test input (STI). The set-up is now complete for the engine off test. Turn the ignition switch on and count the needle sweeps. During the engine off self-test, there will be two groups of codes and three events.

1. The first event is a series of clicks heard in the engine compartment. During this series of clicks, the ECA is taking readings from all the sensors that are operating with the engine off. It will also be switching actuators on and off to test their operation.

2. The second event is the fast codes. These will be seen on the voltmeter as a series of short, fast pulses from 0 to 1 or 2 volts. The fast codes contain all the codes that will be seen later. The STAR tester and other diagnostic tools read these codes. As far as the use of the analog voltmeter is concerned, these pulses have no purpose other than to verify that the Key-On-Engine Off (KOEO) test has been successfully entered.

3. The KOEO on-demand codes are hard codes. These will be a series of needle sweeps from 0 to a full 12 volts. These codes are two-digit codes read in the needle sweeps.
• A Code 21 would be seen as: pulse-pulse/pause/pulse
• A Code 24 would be seen as: pulse-pulse/pause/pulse-pulse-pulse

Each of the on-demand codes will be repeated twice. If there were several codes to be report-

On older EEC-IV TFI applications, the diagnostic test terminal simply hangs from one of the wiring looms in the engine compartment. Note that the test terminal consists of a six-terminal connector and a single-terminal connector.

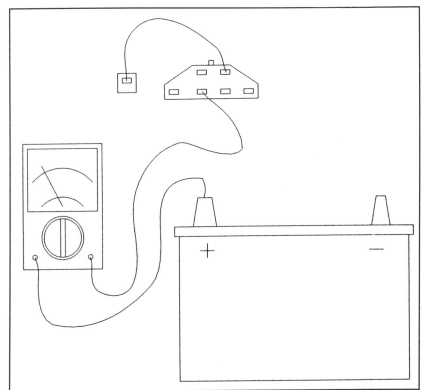

Trouble codes can be obtained with equipment as simple as an analog voltmeter. Connect the red lead of the voltmeter to battery positive. Connect the black lead to the lower left of the center four terminals of the six-terminal test connector. Connect a jumper wire from the upper right of the center four to the single wire connector.

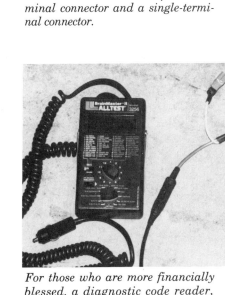

For those who are more financially blessed, a diagnostic code reader, commonly referred to as a scanner, can be used to extract the codes from the EEC-IV computer. In addition to codes relating to the TFI system, information can be obtained about the fuel injection system.

111

ed, they would be reported as: 21-24-21-24

The KOEO on-demand codes represent problems that were detected during the clicking that occurred during the beginning of the test. These are problems that exist now, as the test is being performed.

4. The separator code is next. After the last on-demand code, there will be a five to six second pause followed by a single needle sweep. This single needle sweep is called the separator code. It marks the end of the on-demand codes and the beginning of a set of memory codes. The Ford service manuals refer to this code as a Code 10 (one needle sweep, followed by no needle sweeps).

5. About five or six seconds after the separator code, the memory codes will be pulsed. These will be needle sweeps exactly like those of the KOEO on-demand codes. A code will be set into memory when a major engine sensor circuit is detected to have an open, short, or ground. On late-model EEC-IV applications the MIL (Malfunction Indicator Light) located on the instrument panel will come on when a code is set into memory.

Memory codes represent problems that were detected the last time the engine ran or earlier. These codes represent problems that the ECA detected in sensor or actuator circuits that can have a major effect on the operation of the engine. These problems existed in the past, and for the purpose of interpretation, no longer exist. This can be contrasted with the on-demand codes. The on-demand codes represent problems that exist now, not in the past.

If a serious problem existed in the past, there would be a code stored in memory. If the problem existed in the past and still exists, there will be both an on-demand code and a memory code.

KOER Test

The Key-On-Engine Running (KOER) self-test is a little more complicated to perform than the KOEO test. As in the KOEO test, engine temperature when the KOER is run is critical. The engine temperature for this test must be between 180 and 250deg F. To ensure that the test and the results of the test are accurate, follow this procedure.

1. Disconnect the jumper between the upper right of the center four of the six pocket test connector and the STI pigtail. Start the engine and allow it to run at 2000rpm for two minutes. This is to ensure that the engine coolant temperature and the EGO sensor are heated to their

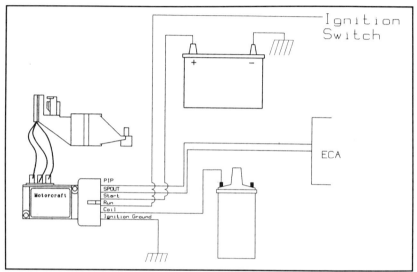

For most applications the TFI system is a distributor type ignition system. Whether the engine is a four-, six-, or eight-cylinder, the ignition system tests and design are the same.

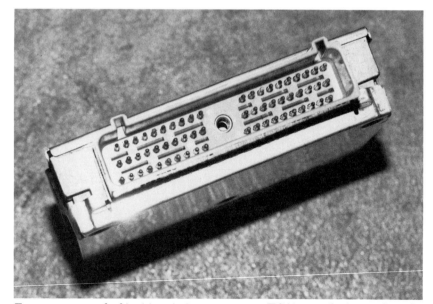

For proper control of ignition timing, the connection between the TFI module and the well-hidden EEC-IV computer (ECA) must be in good condition.

normal operating levels.

2. Shut the engine off, reconnect the jumper, and start the engine. Almost immediately after starting the engine, the engine ID codes will be pulsed. These will be two, three, or four pulses of the voltmeter. Two pulses of the voltmeter mean that the ECA believes it is connected to a four-cylinder engine; three sweeps of the needle means that the ECA believes it is hooked to a six cylinder engine. An eight cylinder is identified by four sweeps of the needle. If the number of pulses does not match the number of cylinders, the wrong ECA has been installed in the vehicle.

Note: Some scanners and engine analyzers may miss the first engine ID pulse. This gives an indication of two less cylinders than the engine really has. If you are using a scanner and get an indication of fewer cylinders than the engine actually has, connect a voltmeter and repeat the test. It is possible that the scanner missed the first pulse.

3. Following the engine ID codes, the ECA will begin to switch actuators on and off, modifying the idle speed. As the ECA does this, it looks for a response from its sensors. An example of this is when the thermactor air divert valve pumps air across the EGO sensor. The ECA expects to see the EGO sensor voltage drop low. If it does not, it will generate codes. During this time the engine rpm will go up, go down, the engine will misfire, and the catalytic converter will stink.

4. During the events that occur during step three, the technician needs to BOOPS—brake-on-off, power steering. The technician must turn the steering wheel one-half turn and touch the brake pedal. If the vehicle has a standard (manual) transmission, depress the brake pedal only. As the technician works these parts, the ECA takes the opportunity to confirm their operation. Although it is not necessary to BOOPS on many applications, if it is not done on a vehicle that requires it, false codes will be generated. If the BOOPS is done on a vehicle that does not require it, no false codes will be generated.

5. After several seconds, the

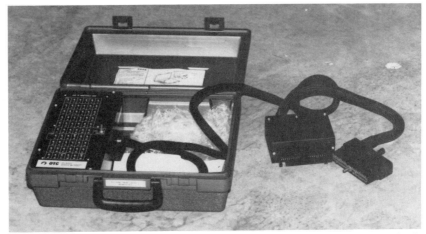

OTC (Owatonna Tool Company) and other automotive tool manufacturers make a tester called a breakout box. This tool, when installed between the ECA and the ECA wiring harness, provides a convenient place to check voltages to and from the ECA. The sixty terminals on the black panel to the right each represent one of the wires connected to the ECA.

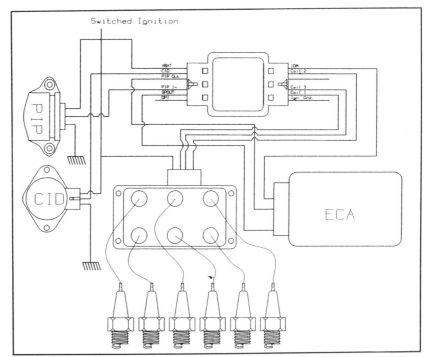

Some later model Ford applications are equipped with the Distributorless Ignition System (DIS), which is essentially the same as the standard TFI system. An additional sensor called a Cylinder Identification (CID) is used to tell the ignition module which cylinder it needs to fire next. There is one ignition coil used for each pair of cylinders. Each coil delivers spark to two cylinders simultaneously.

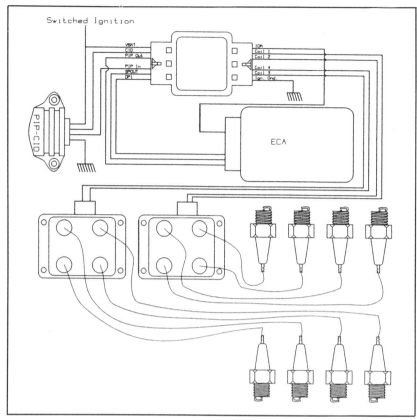

The DIS system used on the 2.3 liter four-cylinder engine uses two spark plugs per cylinder and has two coils having four terminals each.

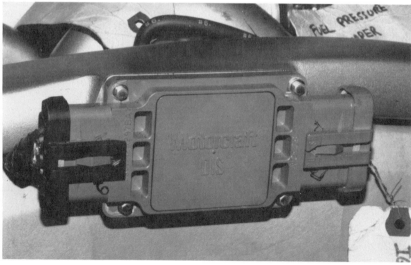

The DIS module has twelve terminals. Starting with the upper right terminal and going counterclockwise, these are: 1. VBAT battery voltage; 2. CID Cylinder Identification; 3. PIP Out the PIP signal from the module to the ECA; 4. PIP In the PIP signal from the PIP sensor to the TFI module; 5. SPOUT the timing control signal from the ECA to the TFI module; 6. DPI Dual Plug Inhibit—shuts down one spark plug on each cylinder while starting to reduce the load on the battery; 7. Ign. Gnd. TFI module ground; 8. Coil 3 fires coil 3 (not cylinder 3); 9. Coil 4 fires coil 4; 10. Coil 2 fires coil 2; 11. Coil 1 fires coil 1; and 12. IDM Ignition Diagnostic Monitor; ECA monitors the condition of the ignition module.

engine will smooth out. At this point, the ECA has advanced the timing to approximately 20deg before initial timing. There will be a pause lasting several seconds followed by a single needle sweep called the Dynamic Response Code (Code 10). Following the single needle sweep, the technician has ten seconds to snap the throttle to the wide open throttle position. The engine speed must exceed 2000rpm. As soon as the target rpm is exceeded, the throttle should be closed.

This is called the goose test. The purpose is to allow the ECA to see changes in the signals from the MAP sensor and the TPS. In addition to the sensor changes, the ECA wants to see that the speed of the engine returns to the proper idle speed following the snap.

6. The goose test is followed almost immediately by fast codes. Like the KOEO fast codes, the KOER fast codes have all the information about what was discovered during the KOER test. The fast codes require a scanner or STAR tester to read.

7. The KOER on-demand codes represent problems that were found as the ECA went through the KOER test. These, like the KOEO on-demand codes, are two-digit codes read in needle sweeps.

The following codes relate to the TFI ignition system.

Code 14

Code 14 indicates a failure in the PIP circuit. This code will never be found in either the KOER or the KOEO on-demand codes. It will only be pulsed out as a memory code. If this memory code is present, it is a result of erratic pulses from the PIP. This condition could be caused by magnetic or radio interference. Be sure that the antenna coax for two-way radios is not routed near the EEC-IV wiring harness.

If there is no reason to believe that there are radio or induction sources, locate the Ignition Diagnostic Monitor (IDM) wire. The IDM on the TFI system (distributor type ignition) is the second wire from the bottom on the six wire connector to the ignition module. For applications equipped with the Distributorless Ignition System (DIS), this is pin number 12. Disconnect the TFI or DIS module. Disconnect the coil or coil pack. Disconnect the 60 pin connector from the ECA. This isolates the IDM wire from the electronic components. Connect an ohmmeter between the IDM wire and ground. The ohmmeter should read infinity. If the ohmmeter reads less than infinity, repair the short to ground in the IDM wire. If the circuit does read infinity, with the ohmmeter still connected, wiggle the IDM wire. This will check for an intermittent short to ground.

If the IDM wire is not shorted to ground, connect the black lead of the ohmmeter to ground. Connect the red lead of the ohmmeter to pin 4, 40, 46, and 60. For each of these pins, the resistance to ground should be infinity. If any show a resistance of less than 10,000 ohms, replace the ECA.

If the resistance to ground for the ECA checks out good, reconnect the coils, ignition module, and ECA. Perform the continuous monitor test on the ignition module wiring harness and the ECA harness. As part of this test, lightly tap on the TFI (or DIS) module and on the ECA to simulate road shock. For information about performing the continuous monitor test (also known as the wiggle test), refer to the end of this chapter.

If the system passes the wiggle test, we have reached an ironic impasse, the problem that generated the Code 14 is intermittent and is not currently presenting itself. Put everything back together and erase the code. Assume that the code was set by a passing radio signal. If

This is the DIS ignition coil used on the 3.0 liter Super High Output (SHO) engine. Other six-cylinder applications are similar.

The CID on the 3.0 liter SHO engine is located on the end of the right bank intake camshaft.

the signal returns, repeat the above procedure. Remember that when a memory is set, the MIL light will illuminate.

Code 18

Code 18 relates to the loss of the tach signal to the ECA. If the Code 18 comes up during the engine running test, it means that the ECA has detected an open circuit in the SPOUT circuit during the engine running self-test on-demand codes. Because the code came up during the on-demand codes, it means that the problem causing the code is a hard fault (an existing problem) and therefore should be easy to trace.

If the Code 18 is in the memory codes, it will be a little more difficult to trace. A memory Code 18 indicates that the IDM signal between the TFI module and the ECA has been lost some time in the past. Because the Code 18 did not come up in on-demand, the problem is intermittent, no longer exists, and may be difficult to locate.

Troubleshooting the KOER on-demand code is fairly simple. Begin by locating the SPOUT connector. This is the connector located in the yellow/green wire that runs from terminal 2 of the TFI module to terminal 36 of the ECA. Disconnect the SPOUT connector to check or set the initial timing. The Code 18 could result if the connector was left out or disconnected during the last tune-up.

If the connector is in place, perform the computed timing test. Connect a timing light. Run the KOER self-test. After the last on-demand code is received, check the timing. The timing should be initial timing (usually 10deg before top dead center on EEC-IV engines) plus 20deg; plus or minus 3deg. If the computed timing test meets specification, and the Code 18 was again received as you prepared for computed timing, you have a real puzzle. Logically, these results are mutually exclusive. Repeat the computed timing test and again watch for the Code 18. If the Code 18 repeats and the computed timing is within specification, the ECA is making erroneous decisions and should be replaced.

If the computed timing is not correct, shut the engine off, disconnect the self-test hookup, disconnect the SPOUT, and check the initial timing. While the initial timing is 10 to 12deg on most EEC-IV applications, it would be wise to check this against the EPA decal located under the hood. If the initial timing is okay, but the computed timing was incorrect, check the power supply voltage to the ECA. With the ECA harness connector connected and the key on but the engine off, measure the voltage between ECA terminal pin 37 and pins 57 and 60. It should be battery voltage. If these voltages are okay, check the voltages between pins 40 and 57, as well as 40 and 60; the voltage should be battery voltage.

If all these voltages are okay, connect a digital tachometer to the yellow/green wire that runs from the ECA pin 36 and the number 2 terminal of the TFI module (this is the SPOUT wire). Start the engine. The tachometer should read an rpm greater than zero. If the tach reads zero, inspect the SPOUT wire for opens or shorts to either ground or voltage. If the wire is okay, replace the ECA. A reading greater than zero on the tachometer means that the ECA is producing the SPOUT (timing control) signal and the TFI module must be impeding the timing control ability of the system.

If the initial timing is incorrect, adjust as necessary and repeat the self-test. If the computed timing is still incorrect, replace the ECA.

Note: If the vehicle being diagnosed is equipped with distributorless ignition, the SPOUT wire is connected to pin 5 of the DIS module.

Memory code 18 means that during an earlier operation of the engine, the ECA detected a loss of the IDM signal to the ECA. This condition could be caused by a shorted or open harness between the ignition module and the ECA, a defective ignition module, or a defective ECA. If the vehicle is equipped with the TFI (distributor ignition system), disconnect the E-core coil and the 60 pin connector from the ECA. With an ohmmeter set on the 200,000 ohm scale (or equivalent), check resistance between the ECA harness pin 4 and the negative terminal of the E-core coil primary. The resistance should be between 20,000 and 24,000 ohms. If the resistance is greater than 24,000 ohms, repair the open circuit in the wire. If it is less than 20,000 ohms, check the resistance between pin 4 and pins 40, 46, and 60 of the ECA harness. If any of the resistances is less than 10,000 ohms, repair the short between the affected wires.

If all the resistances check out okay, reconnect the 60 pin connector to the ECA. Leave the E-core coil disconnected. Again, check the resistance between terminals 4 and 40, 4 and 46, and 4 and 60. The resistance for each test should be greater than 10,000 ohms. If not, replace the ECA. If the resistance is greater than 10,000 ohms, reconnect the E-core coil and perform the Continuous Monitor test (wiggle test).

Enter the wiggle test as described at the end of the chapter. Wiggle the TFI module harness and connector. Pay particular attention to the six wires at the TFI module. Tap on the TFI module. If a fault is indicated, make the appropriate repair. If no fault is indicated, replace the ECA.

If the vehicle you are working on is equipped with the DIS, the test procedure is the same as described above. However, where the instructions say to connect or disconnect the E-core coil, connect the pin 7-12 connector of the DIS module.

Testing the TFI Ignition Module

Other than replacing with a known good unit, there is no accurate way to test an ignition module without special equipment. If the pickup produces a tach reading at the PIP (or PIP OUT terminal in the case of the DIS systems) the problem is wiring or module.

Distributorless Ignition System

The DIS was developed for the 1988 model year. Like the distributor applications, if there is no high-voltage spark, the fault could be in the coil, ignition module, or engine position sensors.

Engine Position Sensors

The DIS system uses Hall Effects sensors to monitor the positions of the camshaft and crankshaft. On the 2.3 liter Rangers, the cam sensor is located in the crankshaft. On all other applications, the cam sensor is located on the camshaft.

Crankshaft (PIP) Sensor

Probe the middle wire at the crankshaft position sensor connector. This wire runs to terminal 4 of the DIS module. The reading should either be greater than 4.5 volts or less than 1.0 volt. Whichever it reads, carefully rotate the crankshaft. Within one-half revolution of the crankshaft, the voltage should change from low (less than 1 volt) to high (greater than 4.5 volts).

If the voltage was high, and remains high, check the continuity of the wire from the PIP to the module. If the wire is good, no shorts or opens, replace the PIP (crankshaft) sensor. If the voltage was low, and remained low, check for 12 volts to any of the terminals. If there are 12 volts, inspect the ground; if the ground is good, replace the sensor. If there were not 12 volts at the sensor, repair the 12 volt power supply wire as needed.

Camshaft (CID) Sensor

Probe the middle wire at the camshaft position sensor connector. This wire runs to terminal 2 of the DIS module. The reading should either be greater than 4.5 volts or less than 1.0 volt. Whichever it reads, carefully rotate the crankshaft. Within one revolution of the crankshaft, the voltage should change from low (less than 1 volt) to high (greater than 4.5 volts).

If the voltage was high, and remains high, check the continuity of the wire from the CID to the module. If the wire is good, no shorts or opens, replace the CID (camshaft) sensor. If the voltage was low, and remained low, check for 12 volts to any of the terminals. If there are 12 volts, inspect the ground; if the ground is good, replace the sensor. If there were not 12 volts at the sensor, repair the 12 volt power supply wire as needed.

Conclusions

If the voltage at terminal 4 of the ignition module switches high and low as the crankshaft is rotated, the crank sensor is good.

If the voltage at terminal 2 of the ignition module switches high and low as the crankshaft is rotated, the cam sensor is good.

There should be 12 volts on the wire connected to the 1 terminal of the ignition module.

There should be continuity to ground through ignition module terminal 7.

If all of the above are true, then replace the ignition module.

Assumption!

The above diagnosis for no spark on DIS assumes that none of the coils are producing a spark. If even one of the coils produces a spark, the problem is most likely a defective coil pack.

Quick Check Tips

1. If there is no spark but the injector pulses, then the lack of spark is a result of a secondary ignition problem such as a bad coil, distributor cap, or rotor. If there is no spark and the injector does not pulse, then the problem is likely a primary ignition problem.

2. A quick and easy test for pulses from DIS cam and crank sensors is to connect a dwell meter to terminals 2 and 4 of the ignition module. If the sensors are good, the tachometer will read greater than zero, if the sensor being tested is bad, the reading will be zero.

General Motors Ignition System 15

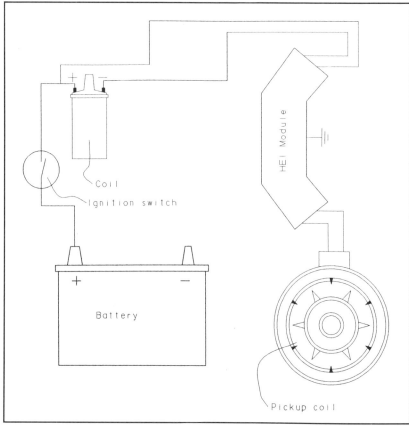

The General Motors HEI ignition system was introduced in 1974. Although by today's standards the system was extremely simple, many mechanics fled from the business because of it.

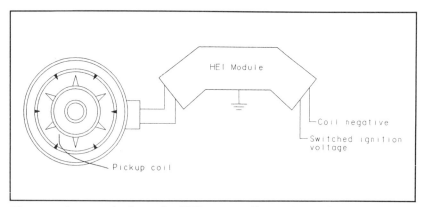

There are four terminals on the HEI ignition module. Two terminals go to the AC pickup coil. A third wire goes to coil negative, and the final wire is connected to switched ignition voltage.

General Motors introduced its first electronic ignition system in 1974. Known as the High Energy Ignition (HEI) system, it features a high current flow through the ignition module and through the primary side of the coil. The higher current in the primary creates an extremely high current and voltage in the ignition secondary. The higher secondary current and voltage permits quality combustion of leaner mixtures.

Primary Ignition Components

Electronic Control Unit

The original HEI system placed the ignition module in the distributor, although in many later model applications the module is located remote from the distributor. There are four terminals on the module. Two of the terminals are connected to the pickup coil. One is connected to ignition power, while the other is connected to the negative side of the ignition coil. The ground for the ignition module is provided through the metal backing.

Pickup Coil

The pickup coil is used to measure distributor shaft rotational speed and produce an AC sine wave with a frequency directly proportional to the speed of the rotation. The primary advantage of the pickup coil over other rotational speed sensors is its simplicity. Consisting of a coil of wire, permanent magnet, and rotating reluctor, there is little that can go wrong with it. Its main disadvantage is its inability to accurately detect low speed rotation. At low rotational speeds, the pickup coil is unable to produce a signal.

How the Pickup Coil Works

A coil of wire sits in a magnetic field created by a permanent magnet. A metal wheel with protruding reluctor teeth rotates through the magnetic field. As it rotates and one of the teeth approaches the magnetic field, the magnetic field is bent toward the approaching tooth. As it is bent, it passes across the coil of wire inducing a voltage. Continuing to rotate, the reluctor tooth drags the magnet field across the coil of wire eventually bending it in the opposite direction. The result is an AC signal.

Testing the Pickup Coil With an Ohmmeter

Most books on troubleshooting electronic ignition systems will describe using an ohmmeter to test a reluctance pickup. Although this is not a totally invalid test, it only tests the coil of wire. Reluctor air gap, condition of the permanent magnet, and adequate rotational speed are not tested with this method.

Disconnect the pickup from the ignition module or vehicle wiring harness leads. Connect the ohmmeter to the pickup coil leads and measure the resistance. Typical ohmmeter readings for a good reluctance pickup coil would be between 500 and 1,500 ohms.

With an Oscilloscope

A much better way to test a reluctance pickup is with an oscilloscope. Connect the scope to the pickup coil leads and rotate the reluctor (crank the engine, rotate the wheel). A series of ripples should appear on the scope. If the line remains flat, there is an open in the coil of wire, and the permanent magnet has been damaged or the air gap between the pickup and the reluctor teeth is too large.

The HEI ignition module is available from a variety of aftermarket sources. Most, like this, are of good quality. A good rule to follow is that if the ignition module you purchase is the cheapest you could find, cheaper than the others, you probably did not get a good module.

It is often forgotten that the ignition switch is part of the primary ignition system. Intermittent dying problems are sometimes traced to the ignition switch.

119

The HEI ignition coil in early and many late-model V-8 applications was located in the distributor cap.

Later model HEI systems use an externally mounted ignition coil.

With an AC Voltmeter

The AC voltmeter is a practical and effective alternative to the oscilloscope. Connect the AC voltmeter in the same manner as described for connecting the oscilloscope. Rotating the reluctor at minimum speed (cranking the engine, rotating a wheel at one revolution per second) would yield between 0.5 and 1.5 volts. If the AC voltmeter does not produce a voltage, there is an open in the coil of wire, and the permanent magnet has been damaged, or the air gap between the pickup and the reluctor teeth is too large.

Notes on Testing

Since the pickup coil is primarily a coil of wire and a permanent magnet, it is prone to intermittent failures with changes in temperature and vibration. If the failure being diagnosed is intermittent, the sensor should be heated and tapped while testing.

Ignition Switch

The ignition switch controls whether current will be available to the coil and ignition module. Unlike the point/condenser ignition system, the HEI system uses the same current path when starting the engine that it uses when the engine is running. There is no ballast resistor in the HEI system.

Secondary Ignition Components
Coil

Depending on the year and engine, the ignition coil may be mounted on the engine, on a fender well, or in the distributor cap. The ignition coil secondary consists of hundreds of windings of very thin wire. When current stops flowing through the primary, the magnetic field created by the primary current flow collapses. This collapsing magnetic field induces several thousand volts in the secondary. This is the voltage that is used to jump the gap of the spark plugs to fire the mixture in the cylinders.

Because of the relatively low current in this high-voltage side of the coil, there are relatively few problems in the secondary side of the coil. The problems that do occur are usually in the form of open circuits.

Coil Wire

The coil secondary output wire carries the high voltage current from the coil to the distributor cap. The coil wire is normally 6 to 12in long and has a resistance of a few thousand ohms. This relatively high resistance helps to reduce the intensity of the radio signal created by the secondary ignition.

All arcing generates a radio signal. In addition to the arcing that occurs at the spark plug, there is also an arc inside the distributor cap between the cap and the rotor. All of the secondary ignition wiring has a high resistance to reduce the affect of this arcing.

The coil wire can suffer from several possible problems. As the coil wire ages its resistance tends to increase. Also, as the wire ages the insulating quality of the jacket decreases, which makes it possible for the high voltages being carried by the wire to penetrate the jacket and arc to ground. Corrosion can also affect the current carrying ability of the wire.

A defective coil wire can result in misfiring, no-start, and poor power. If the ignition coil is located in the distributor cap, there is no coil wire.

Distributor Cap and Rotor

These two components operate as a team. The coil wire delivers the high voltage to the center terminal of the distributor cap. A carbon conductor carries the voltage to the center of the rotor. The rotor will have either a metal or carbon resistive conductor that carries the voltage to the tip of the rotor. The rotor mounts on the top of the distrib-

The center carbon electrode can be removed from the distributor cap and must be replaced along with the distributor cap. The replacement electrode should be found in the box with the new distributor cap.

utor shaft and is driven by the camshaft. As the rotor rotates, it approaches either copper or aluminum conductors on the inside of the distributor cap and arcs to these conductors, which carry the voltage to the spark plug wires.

The distributor cap is prone to cracking, corrosion, and carbon tracking. Carbon tracking occurs when a microscopic crack or piece of dirt provides a current path to ground that is easier than the current path and plug wire. The rotor is subject to corrosion, and perforation. Perforation occurs when the high voltage seeks and finds a ground through the rotor to the distributor shaft.

Routine replacement of the distributor cap and rotor can prevent unforeseen problems. It is not necessary to replace the cap and rotor at each tune-up as many professional technicians recommend, but they should be replaced at every other tune-up. As you read this, however, do not assume that the preceding statement means that your mechanic has been ripping you off for the past 10 years. There is no disservice in a mechanic charg-

ing you $30 or $40 extra at each tune-up to ensure that you have less chance of developing premature problems.

As discussed in Chapter 2, when replacing the distributor cap or rotor, I advise that you replace them as a set. I further ad-

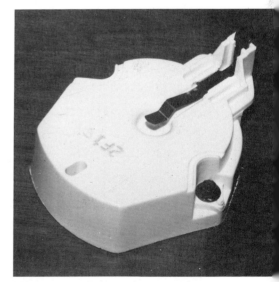

The rotor used on applications that have the coil in the distributor cap have an annoying habit of developing a hole under the spring steel contact for the cap's center electrode. When this happens, the engine will fail to start.

The two wires with the double spade connectors fit into obvious slots in the distributor cap. The black wire with the eyelet connector is the ground wire.

This is the ground connector for the ignition coil. The black wire from the coil should be mated to this connector and the frame of the ignition coil.

vise that they be built by the same name brand manufacturer. I have had situations in the past where a mismatched cap and rotor caused the rotor air gap to be so large that the engine either failed to start or misfired.

Spark Plug Wires

When replacing spark plug wires, the old adage, "You get what you pay for" is especially true. A $50 set of spark plug wires can easily outlast four $12 sets of wires. A marginal plug wire can cause a misfire when the engine is under an extreme load and on a modern fuel injected car can make the engine run rich.

Spark Plugs

Every tune-up includes replacing the spark plugs. There are many brands of spark plugs on the market, some good, some bad. Asking for opinions on which is the best brand of spark plug is like asking a group which is the best soft drink, it is largely a matter of personal opinion. What I have always done, when I had a choice, was to use the brand that the manufacturer installed at the factory. My thinking is that the manufacturer has a vested interest in choosing the spark plug that would provide the best driveability and has the least chance of requiring replacement within 12,000 miles. This method has rarely failed to provide either myself or my customers with good service.

The spark plug consists of a pair of electrodes separated by an air gap of between 0.028 and 0.075in. As the spark from the coil travels down the plug wire seeking ground, it must arc across this air gap. If this arc is exposed to a properly preheated, well atomized mixture of fuel and air, this spark will ignite the fuel.

As the spark plugs arc at a high frequency in high temperatures, the electrode material will slowly vaporize. This causes the gap to widen. The wider the gap, the higher the voltage required to initiate the spark across the gap. Eventually, the voltage required to initiate the spark across the gap will be greater than the coil is capable of generating and a misfire will occur.

Common problems associated with the spark plugs are misfiring and difficulty in starting.

Timing Control

The HEI system on non-computer controlled vehicles uses a standard mechanical/centrifugal advance to compensate for changes in engine rpm. Additionally, the retarding of the ignition when the engine is under a load is accomplished through the vacuum advance diaphragm. The spring-loaded vacuum advance diaphragm uses engine vacuum to hold the pickup coil in an advanced position while the engine is at an idle or not under a load. When the engine comes under a load, engine vacuum drops and the spring in the vacuum advance diaphragm pushes the pickup coil to a retarded position. Retarding the timing during acceleration reduces the possibility of detonation.

Setting the Timing

For most applications, to set the timing reduce the engine rpm to the proper specification. This specification is found on the decal under the hood. Disconnect and plug in the hose to the vacuum advance. Adjust the timing to the proper specification.

Troubleshooting
No Start: Basic Tests

In the mid-seventies there were many mechanics who left the field in favor of the fast food industry or diesel truck repair. The reason for this migration was the feeling, supported by the manufacturers, that it would

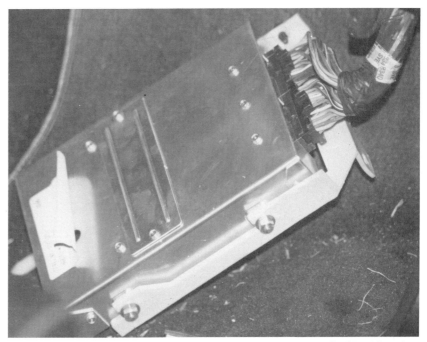

With the advent of tighter emission controls in the late seventies, General Motors found it necessary to begin using computer control of ignition timing.

If the ignition module has five connectors, the ignition system is equipped with Electronic Spark Control (ESC), which retards the ignition timing when the engine pings or detonates.

The ESC module receives a signal from the knock sensor. When a detonation occurs, the ESC module signals the HEI five-terminal module and the timing is retarded.

The seven-terminal ignition module is used with applications that have computer-controlled timing.

This seven-pin module slides into the bottom of the distributor. The two exposed terminals connect directly to the pickup coil.

take a rocket scientist to troubleshoot any problems. That is not true.

After confirming that there is no spark to any of the spark plug wires, remove the distributor cap. Carefully inspect the rotor. The rotor used with the coil-in-the-distributor-cap design has a particularly bad habit of perforating and grounding the spark to the distributor shaft. Inspect the distributor cap for cracks and damage to the center electrode.

Testing the Distributor Pickup

Consisting of only a coil of wire, permanent magnet, and rotating reluctor, there is little that can go wrong with the pickup coil. Its main disadvantage is its inability to accurately detect low speed rotation. At low rotational speeds, the pickup coil is unable to produce a signal.

How the Pickup Coil Works

A coil of wire sits in a magnetic field created by a permanent magnet. A metal wheel with protruding reluctor teeth rotates through the magnetic field. As it rotates and one of the teeth approaches the magnetic field, the magnetic field is bent toward the approaching tooth. As it is bent, it passes across the coil of wire inducing a voltage. Continuing to rotate, the reluctor tooth drags the magnet field across the coil of wire eventually bending it in the opposite direction. The result is an AC signal.

Testing the Pickup Coil With an Ohmmeter

Most books on troubleshooting electronic ignition systems will describe using an ohmmeter to test a reluctance pickup. Although this is not a totally invalid test, it only tests the coil of wire. Reluctor air gap, condition of the permanent magnet, and adequate rotational speed are not tested with this method.

Disconnect the pickup from

the ignition module or vehicle wiring harness leads. Connect the ohmmeter to the pickup coil leads and measure the resistance. Typical ohmmeter readings for a good reluctance pickup coil would be between 500 and 1,500 ohms.

With an Oscilloscope

A much better way to test a reluctance pickup is with an oscilloscope. Connect the scope to the pickup coil leads and rotate the reluctor (crank the engine). A series of ripples should appear on the scope. If the line remains flat, there is an open in the coil of wire, and the permanent magnet has been damaged, or the air gap between the pickup and the reluctor teeth is too large.

With an AC Voltmeter

The AC voltmeter is a practical and effective alternative to the oscilloscope. Connect the AC voltmeter in the same manner as described for connecting the oscilloscope. Rotating the reluctor at minimum speed (cranking the engine, rotating a wheel at one revolution per second) would yield between 0.5 and 1.5 volts. If the AC voltmeter does not produce a voltage, there is an open in the coil of wire, and the permanent magnet has been damaged, or the air gap between the pickup and the reluctor teeth is too large.

Notes on Testing

Since the pickup coil is primarily a coil of wire and a permanent magnet, it is prone to intermittent failures with changes in temperature and vibration. If the failure being diagnosed is intermittent, the sensor should be heated and tapped while testing.

Testing the Module

There is no really scientific way to test an HEI ignition module. One trick that will prove the module is in good enough condition to start the engine is the wet finger test. With the distrib-

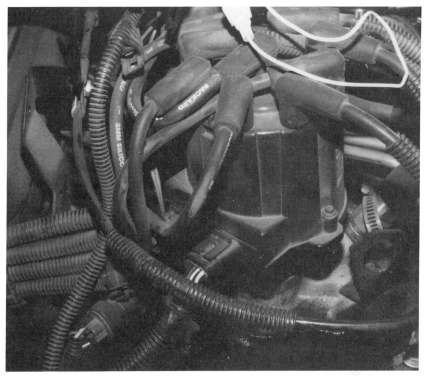

This is a view from the rear engine compartment bulkhead toward the distributor on a 305 (5 liter) engine. This vehicle is a van. Replacing the ignition module on this application is easy. When the same engine is used in the Caprice, the job requires removing the distributor from the engine.

utor cap removed and all wires connected, disconnect the wires that run between the pickup coil and ignition module. Disconnect one of the plug wires from one of the spark plugs, insert a screwdriver in the end of the wire, and hold it a quarter inch from ground. Turn the ignition switch to the run position. Dampen the tip of your index finger and touch the two ignition module terminals that were connected to the pickup coil. If there is a spark between the screwdriver and ground, the module is good. If there is no spark, the ignition module cannot be condemned until all other possible causes of the no-start are eliminated.

Testing the Coil

The resistance between the primary terminals of the coil should be no more than 1.0 ohm. The resistance between the secondary output terminal and the primary should be between 6,000 and 30,000 ohms. Integral coils on applications later than 1975 should have infinite ohms between the secondary and the primary. However, 1980 and later applications should be replaced if there is infinite ohms between both the primary and the tach lead. This is also true of applications later than 1980 with an externally mounted coil.

Dies While Driving Down the Road

If this problem is experienced, the cause is most likely a defective ignition module. Before replacing the ignition module, check the pickup coil for intermittent opens. This is done by tapping on the pickup coil and heating the pickup coil while checking its resistance. The resistance at no point should exceed 1,500 ohms or be less than 500. If the resistance varies outside of these parameters, replace the pickup coil and test drive.

Misfire at Idle

This is almost always never the fault of the HEI system; it is almost always the fault of some component in the secondary side of the ignition system.

Before troubleshooting any misfire, it is essential to verify that the engine is in good condition. A compression test is a good starting point. If the valves are adjustable, be sure that they are properly adjusted.

With a pair of "sissy" pliers, remove and replace one plug wire at a time from the spark plugs. As each plug wire is removed the engine rpm should drop. If one of the cylinders fails to produce as great a drop in rpm as the others, that cylinder is the source of the misfire.

Assuming the cylinder is in good condition and the valves are properly adjusted, remove the spark plug wire for that cylinder and check the resistance. The resistance should be less than 10,000 ohms per volt. If the resistance is correct, replace the spark plug. Unless the spark plugs are very new, replace them all.

Misfire Under a Load

Assuming the engine is in good condition, begin troubleshooting this problem by checking the spark plug gap. If they are gapped properly, replace the spark plugs. Even new spark plugs can misfire under a load.

If replacing the spark plugs does not solve the problem, remove the distributor cap. Inspect the wiring to the points. Frayed wiring can cause an intermittent open circuit as the vacuum advance moves the breaker plate. The intermittent open can cause a misfire.

Lack of Power: Mechanical Timing

There are many things that can cause a lack of power, some related to the ignition system, some not. Begin checking this problem by confirming the engine is in good condition as are the air and fuel filters.

If a lack of power is the result of problems in the ignition system, it is likely the problem is in the timing control system. To test the timing control system, connect a timing light to the engine. Disconnect the vacuum advance and plug in the hose. With the engine at idle speed, check the timing. Now raise the engine speed to 2000 to 2500rpm. If the timing does not advance, the centrifugal advance system is not working. Inspect the distributor weights. If they are free and move easily, replace the weight springs. If the springs are weak, they will allow the timing to advance all the way prematurely, even at idle. If the weights are frozen, use penetrating oil or whatever is necessary to free them. If they are badly corroded, it may be necessary to replace the distributor.

If, or when, the centrifugal advance is working properly, with the engine still at 2000 to 2500rpm, reconnect the vacuum hose to the vacuum advance. When the vacuum hose is reconnected, the timing should advance several degrees.

In the applications that use computer controlled timing, seven wires are connected to the ignition module. Two wires are connected to the AC pickup coil. One is connected to battery positive, while a fourth is connected to the negative terminal of the ignition coil. The black/red wire at the bottom of the above wiring diagram is not attached to the ignition module, but rather is a ground wire that runs from ECM (onboard computer) to the ignition module. The tan/black Bypass wire will have no voltage when the engine is being cranked; when the engine starts the voltage should rise to 5 volts. The purple/white wire is called the HEI reference wire and delivers the engine speed and injector sequencing signal to the ECM. The white wire carries the timing control signal from the ECM to the ignition module.

Lack of Power if Equipped with Electronic Spark Control

In addition to controlling the fuel injection system, the Engine Control Module (ECM) also controls ignition timing. The ECM changes the ignition timing in response to engine rpm, load, and temperature.

Temperature Sensor

In the old days, a thermo-vacuum switch (TVS) was used to apply full manifold vacuum when the engine was cold to the distributor vacuum advance, allowing for full advance and improved cold driveability. When the engine warmed up, the TVS, also known as a Coolant Temperature Override (CTO) Switch, would cut off the manifold vacuum and switch on ported vacuum to the vacuum advance. GM's computerized timing control system eliminates the CTO and uses the coolant temperature sensor input to the ECM in its place.

When the engine is cold, the timing is advanced by the ECM to full advance; when the engine warms up, the idle timing is cut back and total advance is limited.

Tach Input to EST Program of ECM

The old-style distributors had a mechanical system for advancing ignition timing as the engine speed increased. The need to advance the ignition timing as rpm increases comes from the fact that the combustion time is the same regardless of the speed of the engine, about 2 to 3 milliseconds, yet the crankshaft covers more degrees of rotation in the same period of time the higher the rpm of the engine. The ECM on the late-model, fuel injected cars receives a tach signal from the ignition module; as the engine speed increases, the ECM responds to this signal by advancing the timing.

The coolant temperature sensor can affect the ignition timing on applications that are equipped with an ECM. This includes the fuel-injected models 1982 and later, and the engine equipped with electronic carburetors.

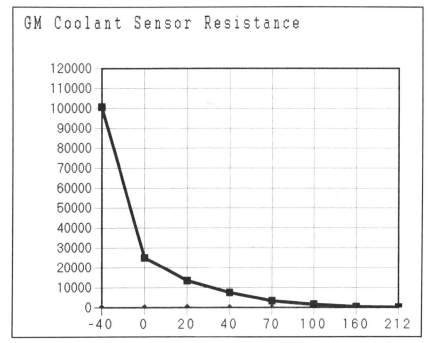

The coolant temperature sensor is a negative temperature coefficient thermistor. As the temperature of the engine rises, the resistance of the thermistor drops. As the resistance of the thermistor decreases, it pulls the voltage on the wire back to the ECM toward zero volts. The ECM is programmed to believe that the lower the voltage on this wire is, the hotter the engine coolant. The hotter the engine coolant, the less timing advance is allowed. Preventing excessive timing advance reduces detonation.

Engine Load

When the engine is under a load, combustion pressures increase; this increases the possibility of detonation. Old technology addressed this problem by taking advantage of the fact that when the engine is under a load, the manifold vacuum decreases. The decreasing manifold vacuum allowed the vacuum advance unit to decrease the amount of advance or, in other words, retard the timing when the engine load increases. New technology uses the ECM's engine load measurement either directly from the MAP sensor for those cars equipped with one, or from the Load Variable from those not equipped with a MAP sensor, to retard the timing when the engine load increases. This accomplishes what the old technology did but with a greater degree of accuracy.

Detonation System

Some GM Electronic Spark Timing (EST) systems incorporate a microphone-like device in the cylinder head or block called a knock sensor. The knock sensor detects vibrations that result from knocking or pinging. Back in the good old days, knocking and pinging were controlled by rich mixtures, leaded gas, and low compression ratios. Today's lean running engines, which run on unleaded gas at gradually increasing compression ratios, have to depend on electronics to guard against detonation.

When a detonation is heard by the knock sensor, the timing retards 3 or 4deg each second until the knock goes away. It then begins to slowly advance the timing until, assuming that engine loads remain steady, the knock sensor begins to hear detonation. It backs the timing up slightly and will remain at that setting until the engine load or throttle position changes.

At first glance, a system to retard timing might seem to work against good performance and power; the opposite is true, however. A detonating engine is one that is wasting power, the air/fuel charge in the combustion chamber is being ignited too soon and is therefore trying to drive the piston back down the cylinder as the crankshaft is trying to push the piston up on the compression stroke, this robs power and wastes fuel. The knock retard system controls detonation, thereby restoring the large part of this lost power and economy to the engine.

These functions are closely tied with the electronic fuel injection system. If there is a problem with these systems in the control of ignition timing, there will be a much more serious problem with the control of the fuel injection system. In fact, typically, the technician or consumer repairman will fix the timing control problem while trying to fix the fuel system problem. For more information about troubleshooting these circuits, consult *How to Repair and Modify Chevrolet Fuel Injection*, also published by Motorbooks International.

Replacing the Secondary Ignition Cables, Cap, and Rotor
Testing and Replacing Spark Plug Wires

There are two ways of testing spark plug wires. The first is with the use of an engine analyz-

Like the Chrysler products, the General Motors applications monitor intake manifold pressure to determine engine load. This is the internal electronics of the manifold absolute pressure sensor. The MAP sensor voltage increases as engine vacuum decreases.

Inside the electronic cell of the MAP sensor there is an electronic strain gauge. A thin silicone diaphragm joins four resistors like a trampoline. As the pressure on the top of the diaphragm increases, the resistors are stretched and their resistance changes. As their resistance changes, the output voltage changes.

er oscilloscope. The second method is with the use of an ohmmeter. Remove each plug wire and with your ohmmeter on the x1000 ohm scale measure their resistance end to end. A good plug wire will have less than 10,000 ohms but greater than 1,000 ohms per foot.

There are only a couple of important things to remember when replacing secondary ignition wires. First, if the plug wires are installed in the incorrect order, a backfire may occur resulting in damage to the air flow meter, air mass meter, or the rubber tube that connects it to the throttle assembly. The original equipment plug wires have numbers on them indicating to which cylinder they should be connected. Aftermarket or replacement plug wires may not have these numbers. 3M and other companies make adhesive numbers that you can attach to insure proper reinstallation.

Replacing the Coil-in-Cap Distributor Cap

As early as 1974, GM began using their high energy ignition system with the coil-in-the-cap distributor. Several V-6 and V-8 fuel injected applications still made use of this distributor through the latter half of the eighties.

When replacing the distributor cap, the coil must be removed from the old cap (unless it is also being replaced) and installed in the new cap. Take care that the three little components supplied with the new cap, the rubber washer, the spring, and the carbon nib, are installed in the proper sequence.

The rotor on these distributors is installed with a pair of screws. A square and round fail-safe stud prevent installing it backwards.

Replacing the Coil-Not-in-the-Cap Distributor Cap

This distributor cap is more conventional in design. Replacement consists of simply releasing the attachment screws or clips and making sure that the plug wires are reinstalled in the correct order.

Whichever type of distributor you have on your car, replace the distributor cap and rotor together and use the same brand. Pairing caps and rotors of two different manufacturers is not a good idea as this can result in the incorrect rotor air gap. Excessive rotor air gap can cause excessively high spark initiation voltage and can result in incomplete combustion.

Replacing the Spark Plugs
Removal

With the engine cold, unscrew the spark plugs two or three turns. If a spark plug is difficult to unscrew, it might have dirt or grit on the threads. Put a couple of drops of oil on the exposed threads and allow it to soak in for a few minutes. Screw the plug back in, then out several times, add a little more each time until the plug is removed. Using compressed air or a solvent soaked brush, clean the area around each plug to remove any dirt or foreign objects that might have fallen into the cylinder when you removed the plugs. Inspection of the removed spark plugs can tell you much about the running condition of the engine. Carbon deposits, ash formations, oil fouling, soot, and other conditions can be indicative of engine or injection system problems.

Installation

Before installing the new spark plugs, be sure to check the gap. Although some spark plug manufacturers make an effort to

This photograph may be the only time most people see a knock sensor, which is usually well hidden. The knock sensor could be compared to a microphone. When a detonation is heard, a signal travels to the ESC module.

pre-gap their plugs, they can be unintentionally re-gapped in shipping. Inspect the threads in the cylinder head to insure that they are clean and undamaged. Also, check the mating surface where the plug contacts the head, it should be clean and free of burrs. Start the new spark plug and screw it in a few turns by hand. Continue to screw the plug in either by hand or with a socket and ratchet until it contacts the mating surface firmly. To avoid overtightening, use a torque wrench and tighten a tapered seat plug to 26 ft/lb in a cast iron head or 21 ft/lb in an aluminum head. In the real world, a torque wrench is seldom used when installing spark plugs so use this rule of thumb: firm contact plus 15deg of rotation.

Checking and Testing the Timing Control System
Checking Timing

On 1982 and later fuel injected GMs, the ignition timing is controlled by the fuel injection computer or ECM. There are four wires involved in the control of ignition timing. All of them run from the distributor to the ECM. A purple wire with a white tracer sends a tach signal from the ignition module to the ECM. The ECM then modifies the signal by changing its duty cycle and returns it to the ignition module through a white wire. In the ignition module, this signal is used to control ignition timing. There is also a tan and black wire that signals the ignition module to ignore the signal on the white wire while the engine is being cranked. During cranking, there is 0 volt on the tan and black wire; after the engine starts, the ECM puts 5 volts on the wire, which notifies the ignition module to use the signal on the white wire to control ignition timing. The fourth wire is a black wire that usually has a red tracer; this wire is simply a common ground between the ECM and the ignition module.

To check the ignition timing, follow the instructions on the EPA decal found under the hood. These instructions will be a variation on one of the following:

1. Disconnect the tan and black wire located somewhere in the engine compartment. This will turn on the check engine light and set a trouble code 42 as soon as the engine is started. Start the engine, adjust the rpm according to the specification on the EPA decal, check or set the initial timing, then reconnect the tan and black wire and clear the code.

2. Place a jumper wire between terminals A and B of the ALCL connector. Check or adjust the timing to manufacturer's specifications. This method will set no code.

3. Disconnect the four wire connector located near the distributor. Adjust timing as described in step one. This will also set a Code 42; do not forget to clear it when you are done. (See Diagnostic Tips in this chapter for procedure to clear codes.)

After preparing to check initial timing through one of the methods described above, use a timing light to check the timing. Adjustment of the timing is still done through loosening and rotating the distributor. On some applications, the timing advance system will not begin to function again until the engine has been shut off and restarted.

If you are working on a car equipped with a Distributorless Ignition System (DIS), you will not find an initial timing specification or adjustment procedure. As discussed later in this book, the timing is controlled entirely by the computer on these.

Testing the Timing Control System

With the car in park and the parking brake set, rev the engine while monitoring the timing with a timing light. The timing should advance considerably. At this point, you are not so much concerned with how much it advances as you are with the fact that it does advance. A DIS can be tested for timing advance by making your own marks on the harmonic balancer and the timing chain cover. Even though there is no adjustment for initial timing, proper advance of timing is still essential for good acceleration and power. If the timing does not advance, the possibilities are:

1. You did not cycle the ignition key off then restart after checking initial timing.

2. Some applications require that the transmission be in forward or reverse gear before the ECM advances timing. Timing advance on these applications can only be checked when the drive wheels are elevated or on a chassis dyno.

3. There are some applications that require an indicated vehicle speed before timing advance can occur. Again, timing advance on these cars can only be tested with the drive wheels elevated or on a chassis dyno.

Failure of the EST to Advance Timing

If you have checked the possibilities mentioned above and the timing still does not change in response to changes in rpm, it is time to grab the voltmeter and wiring diagram.

Testing the Spark Timing Out Wire

There is a white wire referred to as spark timing out that the ECM uses to control the switching of the ignition module to control timing. There should be a pulse with a varying duty cycle on this white wire. If the signal is present at the ignition module but there is no timing advance occurring, then the ignition module is defective.

Diagnostic Tips

To test for the presence and proper function of the signal on the white wire, attach a dwell meter (or, if available, a duty cy-

cle meter) to the white wire as close to the ignition module as is practical. As you rev the engine, the dwell shown on the meter should change.
• If the dwell changes but the timing does not, check the connections at the ignition module.
• If the connections are clean and tight, then replace the ignition module.
• If there is no change in the dwell reading (it reads either 0 or 90deg on the four cylinder scale), then the problem is either in the ECM or the tan/black EST bypass wire.

Checking EST Bypass (tan/black wire)

Check for 5 volts on the EST Bypass wire as close to the ignition module as possible. One convenient place to test would be at the under-the-hood connector that is disconnected to check initial timing.
• If there are 5 volts at this point, check for 5 volts at the point where the tan/black wire connects to the ignition module.
• If there are not 5 volts present at the ignition module on the tan/black wire but the engine runs, the problem could be one of the following:
—an open or ground in the tan/black wire
—a bad connection for the tan/black wire at the ECM
—a grounded (defective) module
—a bad ECM
• If there is 5 volts at the ignition module connection but no signal on the white wire, then there is a bad connection of the tan/black wire at the ignition module, or the ignition module is defective.

Testing for the Dead Hole (Distributor Type Ignition)

When a late-model fuel injected engine has a cylinder with a misfire, the effects can go far beyond a mere rough idle or loss of power. A cylinder that is still pulling in air but not burning that air will be pumping un-

The knock sensor sends its signal to the ESC module. On most applications, this module is located somewhere under the hood. In the Chevro-
let vans of the late eighties, it was mounted on the side of the ECM that was under the driver's seat.

burned oxygen passed the oxygen sensor. This confuses the ECM, making it believe that the engine is running lean. The ECM responds by enriching the mixture, and the gas mileage deteriorates dramatically.

There are several effective methods to isolate a dead cylinder. All of these methods measure the power produced in each cylinder by killing them one at a time with the engine running a little above curb idle.

Back in the good old days, we used to take a test light, ground the alligator clip, and pierce through the insulation boot at the distributor cap end of the plug wire. This would ground out the spark for one cylinder, and an rpm drop would be noted. The greater the rpm drop, the more power that cylinder was contributing to the operation of the engine. Actually, this is a valid testing procedure; however, piercing the insulation boot is only asking for more problems than you started with.

Another previously used method of performing a cylinder balance was to isolate the dead hole by pulling off one plug wire at a time and noting the rpm drops. The problem with this method is that you run the risk of damaging either yourself or the ignition module with a high-voltage spark. So let's explore some valid alternatives.

1. Cylinder Inhibit Tester. Several tool companies produce a cylinder shorting tach/dwell meter. These devices electronically disable one cylinder at a time while displaying rpm. Engine speed drops can be noted. Unfortunately, these testers can cost $500 or more and most will not do a cylinder balance on a distributorless engine.

2. A Good Shoestring Technique. Another valid method does the old test light technique one better. Cut a piece of 1/8in vacuum hose into four, six, or eight sections, each about 1in long. With the engine shut off and one at a time, so as not to

confuse the firing order, remove a plug wire from the distributor cap, insert a segment into the plug wire tower of the cap, and set the plug wire back on top of the hose. When you have installed all the segments, start the engine. Touching the vacuum hose conductors with a grounded test light will kill the cylinder so that you can note rpm drop. Again the cylinder with the smallest drop in rpm is the weakest cylinder.

Whichever method you use, follow this procedure for the best results:

1. Adjust the engine speed to 1200 to 1400rpm by blocking the throttle open. Do not attempt to hold the throttle by hand, you will not be steady enough.

2. Electrically disconnect the IAC motor to prevent its affecting the idle speed.

3. Disconnect the oxygen sensor to prevent it from altering the air/fuel ratio to compensate for the dead cylinder.

4. Perform the cylinder kill test; rpm drop should be fairly equal between cylinders. Any cylinder that has a considerably lower rpm drop than the rest is weak. Proceed to step five.

5. Introduce a little propane into the intake, just enough to provide the highest rpm. Repeat the cylinder kill test. If the rpm drop from the weak cylinder tends to equalize with the rest, then you have a vacuum leak to track down. If there is no significant change in the power output from the weak cylinder, proceed to step six.

6. Open the throttle until the engine speed is about 1800 to 2000rpm. Repeat the cylinder kill test adding propane. If the rpm drops are now equal, then the most likely problem is that the EGR valve is allowing too much exhaust gases to enter the intake at low engine speeds. Remove the EGR valve and inspect for excessive carbon buildup and proper seating. If the rpm drop on the cylinder in question remains low, then the problem is most likely engine mechanical. Proceed to step seven.

7. Perform both a wet and dry compression test. If you have low dry compression and low wet compression, then the problem is a bad valve or valve seat. If the dry compression is bad but the wet compression is good, then the problem is the piston rings. If the compression is good, both dry and wet, then the problem is in the valve train, such as the camshaft, lifters, or pushrods.

8. After testing is completed, reconnect anything that was disconnected or removed for testing.

Testing for the Dead Hole (Distributorless Ignition)

If you were to buy a tester capable of doing a cylinder balance on an engine equipped with a DIS, you would need to plan on spending thousands of dollars. However, the vacuum hose trick described above will work nicely.

Be sure to disconnect the IAC and oxygen sensor and stabilize the rpm at 1200 to 1400 for the first balance test. The rest of the testing procedure is exactly what it was for the distributor type ignition.

Tune-Up Related Serial Data

Those who want to dig a little deeper into the thoughts and decisions of the ECM will need to purchase a scanner. The scanner plugs into the ALCL connector under the dash and displays the information that the ECM is receiving from its sensors, a little about the decisions it is making from that information, and what it is commanding its various actuators to do. A good scanner can cost from a few hundred to several thousand dollars. It is not a purchase to be taken lightly, but through the scanner it is possible to diagnose driveability problems and determine what corrective actions the ECM has

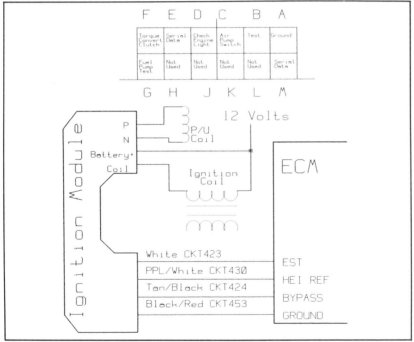

The ignition timing can be set two ways. Follow the instructions under the hood for the method advised for the application being worked on. On many applications, place a jumper wire between terminals A and B of the twelve-socket ALDL (also called the ALCL) connector with the engine idling. On most applications, the ALDL connector is the diagnostic connector located under the driver's side of the dash.

made in its fuel injection or timing control program.

1. INT. The Integrator (INT) is one of the key pieces of serial data information. First, let us refer to an imaginary dictionary where the GM engineers have translated phrases into numbers. If we looked up the phrase "normal for the Integrator" we would find it translated "128 ± 6." With the engine warmed up and in closed loop, request that the scanner display the INT function; 122 to 134 is a normal reading. If the number displayed is greater than 134, the fuel injection system is responding to a request for more fuel. The oxygen sensor must be sensing a high oxygen content in the exhaust, which it interprets as a lean condition. The ECM assumes the engine is running lean and enriches the mixture. Keep in mind that a lean air/fuel ratio is only one of the possible causes of a high oxygen content in the exhaust.

If the Integrator number is less than 122, the engine is receiving extra fuel from somewhere. The Integrator is therefore compensating by decreasing the amount of fuel that the injector(s) is (are) delivering to the engine.

The Integrator changes in direct response to changes in exhaust oxygen content. Whenever the engine is in open loop operation, the Integrator number will be fixed at 128. Once the engine enters closed loop operation, the Integrator will be constantly changing to correct minor errors in air/fuel ratio.

If the Integrator number is high, look for a source of extra oxygen in the exhaust such as:
• Vacuum leak
• Restricted injector(s)
• Cracked exhaust manifold
• Defective air pump upstream/downstream switching valve
• Mechanical engine problems
• Low fuel pressure
• Defect in the ignition (plugs, cap, rotor, plug wires)

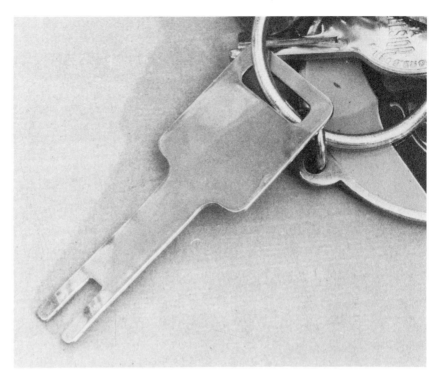

This key is carried by all self-respecting General Motors technicians. It provides an easy method for jumping terminal A to terminal B of the ALDL to set the ignition timing. Additionally, when terminals A and B are bridged with the key on but the engine off, the trouble codes can be retrieved from the ECM.

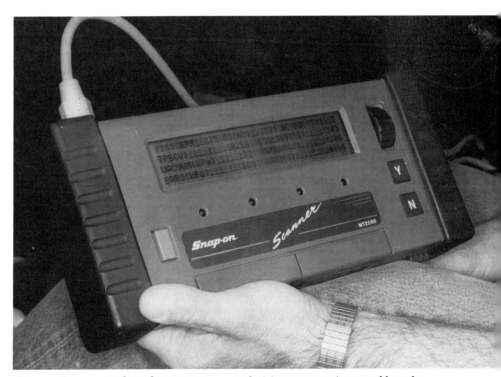

Diagnostic scanners have become one of the most important tools of automotive technicians. With this little box, or one from a plethora of competitors, technicians can retrieve trouble codes and monitor the thought processes of the computer.

If the Integrator number is low, look for a source of extra fuel such as:
• Defective evaporative canister purge system
• Leaking injector
• Contaminated crankcase
• High fuel pressure
• Leaking cold start injector (where applicable)

2. BLM. The Block Multiplier is a long-term version on the Integrator that is stored in a memory. The compensating that it does is for longer term trends in air/fuel ratio. Because the Block Learn is stored in a memory, it even has the ability to compensate for air/fuel ratio problems while the engine is in open loop, providing the problem was detected the last time that the engine ran in closed loop. Just like the Integrator normal is 128 ± 6, a higher number indicates that the ECM is adding more fuel to compensate for a perceived lean condition. A lower number indicates the ECM is compensating for a perceived rich condition. Any abnormal reading in the Block Learn indicates the perceived need to adjust air/fuel ratio has been there for a minimum of several seconds. Several seconds may not seem like a long-term situation, but do not forget we are dealing with a computer.

Always keep in mind when troubleshooting that rich and lean on any modern oxygen feedback fuel injection system are determined by the oxygen content of the exhaust gases, not by fuel entering the engine.

3. O_2. As displayed on the scanner, oxygen sensor voltage should be constantly switching from a low (100 to 300 millivolts) to a high (600 to 900 millivolts) voltage and from a high to a low voltage. Should the oxygen sensor voltage be mostly on the high side, then the engine is running a little rich. Do not become concerned if the oxygen sensor voltage and the Integrator/Block Learn disagree at times. These three functions respond at different rates.

4. Crosscounts. This reading indicates the number of times the oxygen sensor voltage has switched or crossed from a lean to a rich reading, or from a rich to a lean reading. A minimum of four between scan samples on a regular basis is good. The number could be as high as 30 or 40.

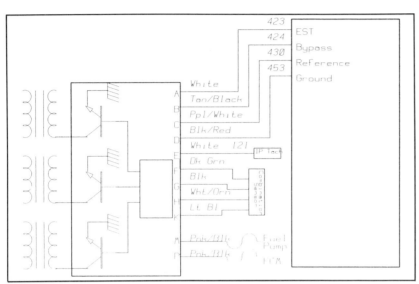

The original General Motors distributorless ignition system was introduced by the Buick division. Called the Computer Controlled Coil Ignition (C_3I), the system used three coils to fire six cylinders.

The C_3I ignition system uses a Hall Effects pickup to monitor crankshaft and camshaft speed. This sensor is from a 3.0 liter Oldsmobile engine. The cam and crank sensor are both located behind the harmonic balancer. The 3800 and other sequentially injected engines use a separate cam sensor on the camshaft.

5. O₂ Status. The oxygen sensor status indicates whether the general trend of the oxygen sensor is to be read rich or lean. It should change back and forth slowly, maybe every 10 to 15 seconds.

6. TPS Volts. During a scan test for tune-up purposes, we are only concerned with the closed throttle voltage that was discussed in the minimum air section of this chapter.

7. GPS Air/Flow. The Grams Per Second (GPS) air/flow relates only to Port Fuel and Tuned Port applications. This is a measurement of the amount of air sensed entering the engine through the MAF. At an idle, this should be about 7 GPS. If the reading is low, then it could be an indication of false air entering the engine. False air can be found by shutting the engine off and stuffing a rag over the end of the MAF. Blow compressed air at about 5 to 10 PSI into the intake manifold. Using a spray bottle filled with soapy water, spray around the rubber connector boot between the MAF and the throttle assembly. Wherever bubbles rise, false air exists. False air can also come in the form of a vacuum leak (after the throttle plates by definition), so spray the entire intake system with the soapy water. Make repairs as needed.

8. Coolant. This is a reading of the engine coolant temperature read in either Celsius or Fahrenheit or both, depending on the scanner.

9. RPM. Engine speed.

10. IAC. This is the position of the idle air control (IAC) motor. There are 256 positions that this motor could be in. The lowest number is zero, the highest is 255. The lower the number, the slower the ECM is trying to make the engine run (7 to 20 are typical at idle for most cars. If the number is at or close to zero, this could indicate a vacuum leak or incorrectly adjusted minimum air.

Serial Data in Tune-ups

Examples of serial data information related to tune-up. These readings are about normal:

INT	128
BLM	128
O₂	100-900mv
Crosscounts	4-16
O₂ Status	RICH
TPS Volts	0.50
GPS Air/Flow	7
Coolant	95C
RPM	800
IAC	15

These readings indicate a problem:

INT	155
BLM	128
O₂	100-400mv
Crosscounts	1
O₂ Status	Lean
TPS Volts	0.50
GPS Air/Flow	3
Coolant	95C
RPM	775
IAC	0

Look at the Integrator reading on the second set of data. It is 155, which indicates that the ECM is adding fuel to compensate for what the oxygen sensor believes is a lean exhaust condition. The normal BLM indicates that this is either a relatively new problem or an intermittent problem that has occurred very recently, perhaps within the past minute. The O₂ does not seem to be indicating rich; a fact that is further confirmed by the low number of crosscounts. The TPS is reading normal, but the GPS air/flow is low. Lean exhaust, high Integrator, and low GPS point toward false air or a vacuum leak.

INT	128
BLM	85
O₂	100-900mv
Crosscounts	8
O₂ Status	Rich
TPS Volts	0.50
GPS Air/Flow	8
Coolant	95C
RPM	800
IAC	13

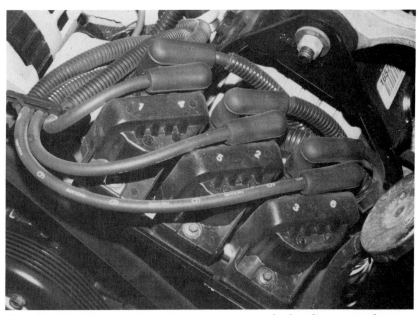

In a distributorless ignition system such as C₃I, each coil sends a spark to two spark plugs simultaneously. In this application, coils 1 and 4 fire together, as do 2 and 5, and 3 and 6. When the coil fires, both spark plugs fire. One cylinder is on the compression stroke heading toward power and one is on valve overlap between exhaust and intake. On all of the General Motors distributorless ignition systems, the coils are mounted on top of the ignition module.

The most noteworthy part of this example is the BLM. It indicates that the engine has been running rich for a while, yet the BLM has fully compensated for the condition and no longer requires the assistance of the INT. The fact that the O_2 voltages are crossing the full range further confirms this. The O_2 status is currently indicating rich but at any moment could swing lean. GPS air/flow is good as is TPS voltage.

For more in-depth information on serial data interpretation, refer to the troubleshooting section.

Summary of Tune-Up Procedures

1. Replace spark plugs
2. Replace air filter
3. Replace fuel filter
4. Inspect distributor cap
5. Inspect distributor rotor
6. Test plug wires
7. Check and/or set initial timing
8. Check for timing advance
9. Clean the mass air flow sensor
10. Clean throttle body coking
11. Check/adjust minimum air
12. Check air/fuel ratio control

Testing the Ignition System for No Starts, Beyond Basic Tests
Distributor Type Ignition

1. In order to test for spark, insert a screwdriver into the spark end of a plug wire and hold the screwdriver a quarter inch from a good ground. Crank the engine and check for spark.

If there is a spark, remove one of the spark plugs, insert it into the plug wire, place the plug on a good ground, such as the intake manifold, and crank the engine. If there is no spark, proceed to step two.

If the spark plug sparks and does not appear worn, proceed to the "No-start, Fuel Diagnosis" section below. If there is no spark at the spark plug, replace the spark plugs.

2. If there is no spark from the spark plug wire, probe the negative terminal of the ignition coil with a test light. If you are not sure which side of the coil is negative, insert it first on one side, crank the engine, then on the other side. One side of the coil should have a steady supply of switched ignition voltage while the other side should flash on and off as the engine is cranked. If neither has power, then you have an open wire in the voltage supply to the coil. If both sides have steady power, then the problem is the primary ignition's control of the coil. In the old days, the first thing to look at would have been the points, today we have to look at that which replaces the points: the ignition module, pickup coil, and the interconnecting wiring.

3. Testing the pickup coil. The variable reluctance transducer (VRT) is the standard type that GM has used since the debut of electronic ignition in 1975. This pickup produces an AC signal. Disconnect the connector at the pickup and connect an AC voltmeter. Be sure that nothing will get tangled or damaged when the engine is cranked. Crank the engine and watch the voltmeter. If it measures a voltage of at least 0.5 volt AC, then the pickup is good, check the wiring and go to step four.

4. Testing the ignition module. Other than replacing the module with a known good unit, there is no accurate way to test an ignition module without special equipment. There is a trick of the trade that will confirm that a module is probably good but will not determine if it is bad. Remove the distributor cap, or access the pickup coil terminals (P and N) of the ignition module whatever way is necessary. Lick the tip of your index finger and touch the P and N terminals simultaneously. Use caution because if the ignition module is good, the ignition coil

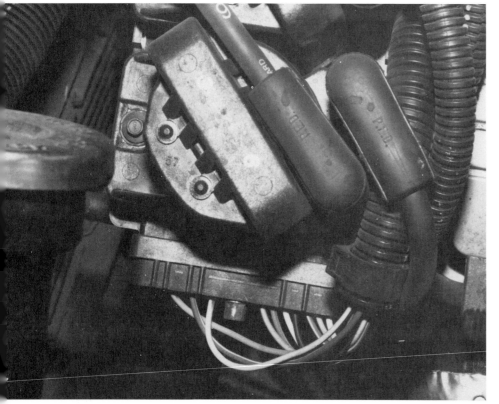

The connector on the end of the ignition module provides a convenient place to test the signals going to and from the ignition module and ECM.

will discharge up to 50,000 volts. Although the current from this discharge is so low that it cannot seriously injure a healthy person, the voltage is high enough to startle you into injuring yourself.

Conclusions

If the pickup produces 0.5 AC minimum, the problem is wiring or module.

If the module triggers through the wet touch test, it is not the module. But if it does not trigger, the module might still be good. If it triggers the coil, carefully inspect the wiring harness; if it does not trigger, inspect the wiring harness and replace the module if no problem is found in the harness.

Computer Controlled Coil Ignition

The Computer Controlled Coil Ignition (C_3I) origination system was developed for the Buick/Olds/Cadillac divisions of GM. This ignition system may be found on GM V-6 applications. Like the distributor applications, if there is no high-voltage spark the fault could be in the coil, ignition module, or engine position sensors.

1. Engine Position Sensors. The C_3I system uses Hall Effects sensors to monitor the positions of the camshaft and crankshaft.

• Crankshaft. Probe the black/light green wire at the crankshaft position sensor connector. This wire runs from terminal H of the ignition module to terminal A of the crankshaft position sensor and provides power for the sensor. The reading should be 10 volts or more with the ignition on.

• Probe terminal B on the crankshaft position sensor connector with an ohmmeter. Although there will be some resistance measured, it should not indicate infinite resistance. If it does, there is an open in the wiring harness, or the module is defec-

There is a fuse, usually 20 amps, located under the hood of the passenger cars. This fuse is located in the fuse panel on light trucks and is marked ECMB or ECM2. When this fuse is blown, the engine will not start. In a no-start condition, it should be among the first things checked. This is also the fuse that is removed to erase the trouble codes.

tive. Measure the resistance in the black/yellow wire that runs from terminal B to terminal G of the ignition module. The resistance should be very close to zero. If the resistance is close to infinity, repair the wire. If the wire is good, check the connection at the ignition module; if the connection is good, replace the module.

Check the voltage with the key on at terminal F of the ignition module, a blue/white wire. The voltmeter should read either close to 0 volt or over 5 volts. Whichever voltage it reads when you rotate the crankshaft, the voltage reading should change. If it was 5 it should drop to 0, if it was 0 it should rise to 5.

• Camshaft. With the key on, there should be 10+ volts in the yellow wire running from terminal N of the ignition module to terminal C of the cam sensor.

Test for this at the cam sensor.

The ground for the sensor is a black/pink wire running from terminal B of the sensor to terminal L of the ignition module. Measuring with an ohmmeter from the B connection to ground should read resistance but not infinity. Like with the crank sensor, an infinity reading means that you need to check the wiring and connections. If they check out good, replace the ignition module.

Check the brown/white wire that runs from the A terminal of the cam sensor to the K terminal of the ignition module. This wire carries camshaft position information to the module in order to sequence the injectors. The voltage should read either 5+ or 0. Rotate the crankshaft. If the voltage was low, it should go high; if the voltage was high, it should go low. If it does not, replace the crank sensor.

Conclusions

If the voltage at terminal F of the ignition module switches high and low as the crankshaft is rotated, the crank sensor is good.

If the voltage at terminal K of the ignition module switches high and low as the crankshaft is rotated, the cam sensor is good.

There should be 12 volts on the pink/black wires connected to the P and M terminals of the ignition module.

There should be continuity to ground through ignition module terminals G and L.

If all of the above are true, replace the ignition module.

Note: If the application you are working on is equipped with a four wire "combination" sensor on the crankshaft and no sensor on the camshaft, there will be a blue wire to terminal K and a dark green wire to terminal F. The diagnostic procedure is the same. Also, terminals J, L, and N will be unused.

Direct Ignition System

Chevrolet division's answer to the C_3I system is the Direct Ignition System (D.I.S.). This is used on four-cylinder and six-cylinder applications. Compared to the C_3I system, D.I.S. is simple. Purple and yellow wires carry an AC signal from the crankshaft sensor to the ignition module. On the ignition module, there are three terminals where these wires connect. The purple wire connects to terminal A and should have 0.7+ volts AC on it when the engine is being cranked. The yellow wire attaches to terminal C and should also have 0.7+ volts AC on it when the engine is being cranked. Terminal B of this connector is shielded and should have no signal on it.

There is also a two wire connector on the ignition module. A pink/black wire is connected to terminal B and should have 12 volts on it when the ignition switch is turned on. There is a black/white wire attached to terminal A, this is the ground.

Conclusions

If there are 12 volts with the key on at terminal B of the ignition module and a ground at terminal A of the two pin connector, then check for an AC voltage of at least 0.7 at both A and C of the three pin connector, then replace the ignition module.

Assumption!

The above diagnoses for no-spark on C_3I and D.I.S. assume that none of the coils are producing a spark. If even one of the coils produces a spark, the problem is most likely a defective coil or coils.

Quick Check Tips

1. If there is no spark but the injector pulses, the lack of spark is a result of a secondary ignition problem such as a bad coil, distributor cap, or rotor. If there is no spark and the injector does not pulse, the problem is likely a primary ignition problem.

2. A quick and easy test for pulses from C_3I cam and crank sensors is to connect a dwell meter to terminals F and K of the ignition module. If the sensors are good, the dwell meter will read somewhere between 10 and 80deg on the four cylinder scale. If not, it will read 90 or 0.

No-Start, Fuel Diagnosis

Testing for fuel pump operation. When the key is turned to the run position but the engine is not cranked, the fuel pump will run for about two seconds then shut off. If you can hear the fuel pump run for this few seconds, you know that the fuel pump relay and the ECM's control of the relay is operative.

Connect a fuel pressure gauge to the Schraeder valve on the fuel rail of the PFI and TPFI applications. No such valve is provided on the TBI applications so it will be necessary to T the gauge into the inbound fuel line. Crank the engine for several seconds, the fuel pressure gauge

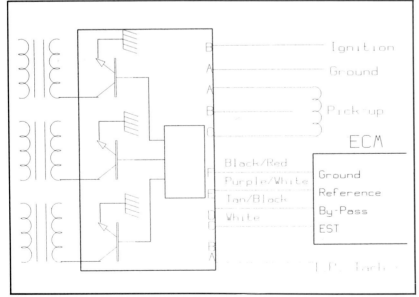

The Chevrolet version of the distributorless ignition system is called D.I.S. (Direct Ignition System). The D.I.S. system is similar to the C_3I system except that an AC pickup coil is used instead of a Hall Effects. The AC pickup slides through the side of the block and uses a toothed wheel on the crankshaft as a reluctor.

should indicate more than 30 but less than 45 PSI for Port and Tuned Port and between 9 and 15 PSI for Throttle Body. This pressure should hold for several minutes to hours slowly bleeding off. If this pressure drops off quickly, then the fuel pump check valve, fuel pressure regulator, or an injector is leaking.

If the pressure is low, run a volume test before and after the fuel filter. Disconnect the fuel filter on the outbound side. Using an approved fuel container, install a hose on the outbound end of the fuel filter and crank the engine for about 15 seconds. The pump should flow a minimum of a pint. If the flow is less, remove the filter and repeat the test. Install a new filter if the flow is good before the filter but reduced after the filter.

If the pressure is low but the flow is adequate, replace the fuel pressure regulator.

If the fuel pressure is correct, check to see if the injectors are opening. There are several good ways to do this. The best way is to use a stethoscope to listen for them to open and close while the engine is being cranked. If a mechanic's style stethoscope is not available, a piece of 1/2in heater hose held near the ear and next to the injector will do almost as well. If the injectors are not clicking and there is no spark, repair the ignition system first. If there is spark but the injectors do not click, check for voltage while cranking at the pink/black wire that supplies voltage for the injectors. If voltage is present, connect a tachometer to the negative side of the injectors.

Cranking the engine should show some sort of a reading (we do not care what the reading is as long as it is not 0). If there is no reading, connect a dwell meter set on the four cylinder scale to the HEI REF input to the ECM. Crank the engine, the dwell meter should read about 45. If the reading remains high or low (90 or 0), check continuity in the purple/white wire from the ignition module to the ECM. On engines equipped with C_3I, it is also necessary for the ECM to receive a signal from the cam sensor. Connect a dwell meter to the black wire that runs from the ignition module to the ECM. Although the reading on this wire will be erratic, we can consider it okay if the reading is anything other than 0 or 90.

If the pink/black wire is carrying 12 volts to the injectors, if there is a pulse on the HEI REF wire, if the wires are good, and if, on the C_3I system, there is a pulse on the black cam sync wire, check for continuity from the injectors to the appropriate ECM terminal (such as D15 or D16 for the 1987 5.7 liter Tuned Port 'Vette).

Purposes of ALCL Terminals

An important thing to note as we look at the purpose of each of the ALCL terminals is that not all applications will have actual connections in each of the twelve terminal pockets. The only two that always have connectors in them are terminals A and B.

A

Terminal A is a ground.

B

Terminal B is the actual self-diagnostic terminal. Grounding terminal B (usually to terminal A) will put the ECM into the field service mode. The field service mode has two forms: Key-On-Engine Off and Key-On-Engine running. During the engine

The Buick and Oldsmobile divisions of General Motors made it easy to find and work on the ignition module and coils. Chevrolet decided to hide them either on the front side of the engine, between the block and the radiator (V-6's), or on the back side of the engine under the intake manifold (four-cylinders).

off field, service mode trouble codes are repeatedly flashed out by the ECM through the check engine light. With the engine running in the Field Service mode, the ECM will display oxygen sensor status through the check engine light. If the light is flashing on and off very fast (about two times per second), that indicates that the ECM is operating in Open Loop. After running for several minutes, the flashing of the light will slow down considerably. At this point the check engine light will be displaying oxygen sensor status. The light on indicates the O2 sensor is detecting a rich exhaust condition; light off indicates that the O2 sensor is detecting a lean exhaust condition.

Placing a 3,900 ohm (3.9k) resistor between terminal B and ground (terminal A) will put the ECM into the back-up mode. This mode activates the limp home system, and the ECM will be using only rpm, throttle position, and coolant temperature to control the injectors. This mode is automatically entered when the ECM detects a major component failure. Placing the 3.9k resistor across terminals A and B tests the ECM's ability to enter this mode.

Placing a 10,000 ohm resistor across terminals A and B will put the ECM into the special mode. This mode will force the ECM into closed loop, ignoring the run timers, and fixes the idle speed at about 1000rpm with a fixed IAC position. The only practical purpose it would serve in troubleshooting the typical driveability problem is if the problem was an unstable idle. Entering the special mode would fix the IAC position if the idle speed continued to fluctuate. Then you would know for sure that the problem was not erratic IAC motor control.

The final diagnostic mode is entered by connecting nothing to terminal B. This is the open, or road test, mode. It is also the mode in which the car is normally driven.

C

Terminal C is found on some fuel injected Chevrolets that use an air pump. This terminal is connected to the wire that the ECM uses to ground; it activates the air switching valve to allow air to be pumped upstream of the O_2. During the warm-up process, the oxygen sensor is preheated by the air pump pumping air across it. Tremendous heat is generated when the CO and HC laden exhaust gases react with the air from the air pump. This is used to hasten the O_2 to a usable temperature. Therefore, during warm-up the voltage at terminal C should be low and, once at operating temperature, the voltage should be high.

A further use for this terminal is to connect a voltmeter to the oxygen sensor. While monitoring the oxygen sensor voltage, ground terminal C. If the oxygen sensor voltage drops as terminal C is grounded, it proves that the oxygen sensor is responsive and that the air switching (upstream/downstream) valve and solenoid work.

D

Quite often the D terminal is missing, but when it is there it is connected to the check engine light. Applying 12 volts to terminal D will test the check engine light for a burned out bulb. Also, connecting terminal D to any actuator circuit will allow the check engine light to be used as a test light.

E

Terminal E may be the most important terminal in the ALCL connector. At this terminal, serial data is transmitted to the scanner. Through this terminal, it is possible to get an insight

The ignition module for the four-cylinder application has two prongs for each ignition coil as well as a facility for mounting the coils.

into what the computer is thinking and how it makes decisions. See the Data section for an extensive look at the use of serial data information in troubleshooting.

F

The status of the torque converter clutch can be monitored at terminal F. Anytime the brake pedal is not depressed and the torque converter has not been requested to engage by the ECM, there will be 12 volts at terminal F. The voltage will drop to zero when ECM engages the torque converter clutch or the driver depresses the brake pedal. Depressing the brake pedal removes the power supply from the lock-up solenoid. Once the brake pedal is released, the voltage at F will go to 12 volts, then drop to zero again when the converter clutch is re-engaged.

G

On some fuel injected cars where there is a connector in terminal G, it is possible to energize the fuel pump for testing purposes by applying 12 volts. Other applications have this connector in the engine compartment.

H, J, K, and L

These terminals are not used for the fuel injection system.

M

Some applications with "high speed" computers use this terminal in addition to E for the transmission of serial data.

Trouble Codes

Chevrolet has three types of codes that can be extracted through the check engine light. Hard codes are the only codes that will turn on the check engine light with the engine running in the road test mode. If the check engine light is on while driving, there is at least one hard code stored in the diagnostic RAM.

Soft and indicator codes are obtained by entering the Key-On-Engine Off field service mode through placing a jumper wire between terminals A and B of the ALCL. The codes will be flashed out through the check engine light three times, each beginning with a Code 12, the only indicator code. This identifies that the ECM recognizes the fact that the engine is not running and that the ECM has entered the field service mode. After all the codes have been flashed out three times each, the ECM will again flash out the Code 12 three times, then continue to repeat the codes until the field service mode is exited by removing the jumper.

After retrieving the trouble

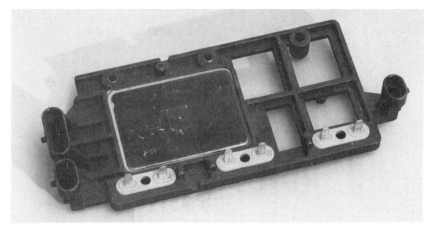

The 3.1 liter ignition module is virtually the same as the four-cylinder version. An additional location for a third coil is also provided.

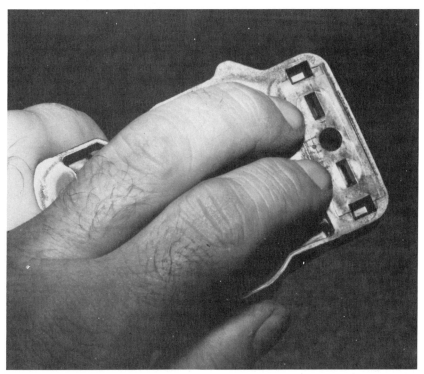

These two openings in the bottom of the ignition coil mount on the spade terminals provided on the ignition module.

141

codes, clear the codes by removing the fuse that powers the diagnostic RAM. This fuse can be found in one of three places. The first is in the fuse panel labeled ECM B or ECM-II. The second is in a fuse holder located in the engine compartment, it can be recognized by the large orange wire going into at least one side of the fuse. The final location is a fusible link coming off the positive terminal of the battery.

After the codes are cleared, reconnect the fuse and start the engine. Run the engine for five to ten minutes until the check engine light comes back on. Retrieve the codes again. As before, the only codes you will get this time will be the hard codes. These codes are the ones you want to diagnose first. Begin by troubleshooting the lowest numbered code first unless there is a 50 code such as 51 or 55.

Soft codes indicate intermittent problems that are not currently detected. These are the codes that did not return after running the engine and pulling the codes a second time.

Code 12

Technically, Code 12 means that the ECM is currently receiving no reference pulses from the ignition system. Since the trouble codes are pulled with the engine not running, the ECM is not receiving distributor reference pulses. This code is not stored in memory and its practical use is to identify that the ECM is in the self-diagnostic mode. This is particularly important when there are no faults; the Code 12 is the only one that will be flashed out.

Code 42

This code is set when the engine is running at greater than 600rpm and there has been no pulse on the white wire from the ECM to the ignition module (CKT 423) for more than 200ms. (At 2000rpm, 200ms represents about 27 spark plug firings.) Code 42 can also be set if the voltage on the tan/black wire (CKT 424) remains low (near zero) or goes low while the car is being driven. The ignition module is designed to ground out the EST signal that travels down the white wire (CKT 423) unless there are 5 volts on the tan/black wire. This locks the timing at a fixed position while the engine is being started.

Common causes of a Code 42 include bad connections, frayed or broken wires, and rubbed through insulation on either the tan/black, white or purple/white wire (CKT 430) running between the ignition module and the ECM.

Code 43

This code relates to the detonation sensor system. GM uses two types of detonation (knock) sensors. The first type dates back to the seventies and uses an interface module between the sensor and the ECM. As of this writing, this is the most common style being used on the Tuned Port as well as most of the PFI and TBI applications. The interface module, known as an Electronic Spark Control (ESC) module has five pins but only four of them are used.

Pin A: Unused

Pin B: CKT 439 is a Pink/Black wire carrying ignition power to both the ESC module and the ECM

Pin C: CKT 485, a Black wire, carries the ESC signal to ECM

Pin D: CKT 486 is a Brown wire which is grounded to the engine to provide a power ground for the ESC module

Pin E: CKT 496 is the White wire which carries the knock sensor's signal to the ESC

On the cars that use the ESC module, Code 43 will be set when the ECM sees the voltage on terminal B7, the black ESC signal wire, has dropped low for more than five seconds with the engine running. A Code 43 might also be set if the ESC system fails the functional test. The ECM performs the functional test once per start-up. When the coolant temperature reaches about 194deg F (95deg C) and

This is a coil from a Type II coil pack. The type two coils can be replaced one at a time. The Type I coil pack design requires that all the coils be replaced together.

the throttle is opened to nearly wide open throttle, the ECM will begin to advance the timing, deliberately allowing the engine to ping in order to test whether or not the ESC will detect the detonation. If no detonation is detected, then a Code 43 will be set into the ECM's memory.

The second type of detonation sensor system is used on the Generation II 2.8 liter and others. This system does not use an interface module, the knock sensor feeds directly into the ECM. Code 43 will be set if the voltage on the dark blue/white wire connected to ECM terminal YF9 drops below 1.5 volts or rises above 3.5 volts for more than half a second. When troubleshooting this circuit, keep in mind that the ECM maintains a 2.5 volt signal on the wire whenever the engine is not knocking.

Code 42 Troubleshooting

Clear the codes then run the engine to confirm that the condition that caused the code to set still exists. If the code returns, then probe the tan/black Bypass wire with a volt meter at the ECM. If the voltmeter reads 0 volt, disconnect the timing set connector in the tan/black wire and note the voltage. If the voltmeter continues to read 0 volt, inspect the tan/black wire back to the ECM. If the wire is okay, replace the ECM.

If, when you disconnect the set timing connector, the voltage increases to 5 volts (engine still running), inspect the tan/black wire to the ignition module. If the wire is good, replace the ignition module.

When you originally probed the tan/black wire with the engine running, if there were 5 volts on the wire, connect a digital dwell meter to the white EST wire. Note the dwell reading. If the dwell meter reads anything other than 0 or 90 on the four cylinder scale, rev the engine. If the dwell changes, connect a timing light to the engine and note timing changes as the engine is revved. If the dwell and timing changes, replace the ECM. If the dwell meter reads 0 or 90, replace the ignition module.

Code 43 (with ESC module)

After confirming that the condition that caused the code to be generated still exists, check for 12 volts on the pink/black wire, which is connected to the ESC module with the key on. Repair the wire or replace the fuse as necessary.

If the pink/black wire has 12 volts on it, turn the ignition switch off, connect an AC voltmeter to the white wire of terminal E on the ESC module. While closely watching the AC voltmeter, tap on the engine block with a 4oz hammer. Does the voltmeter indicate any reading? If it does, the knock sensor is good. If it does not, replace the knock sensor.

If the knock sensor tests good, connect a digital logic probe to the ESC terminal of the ECM. If it reads digital lo with the engine running, inspect the black wire from the ESC module to the ESC terminal of the ECM. If the wire is okay, replace the ESC module. If there is a digital hi on the ESC terminal of the ECM, tap on the engine block, if the logic probe fails to latch for a digital lo, then replace the ESC module.

Integrated Type Knock Sensor: Disconnect the knock sensor wiring connector. Check voltage on the wire going back to the ECM. If there are 5 volts on this wire, replace the knock sensor. If there is 0 volt, inspect the ECM knock sensor input wire for a short or ground. If the wire is okay, replace the ECM.

Bosch Electronic Ignition System 16

Section 1

This section pertains to cars including the 1977-1986 Audi, BMW, Fiat, Mercedes Benz, Peugeot 604, Porsche, Saab, Volvo, and 1986 Yugo.

This system has an ignition module with six terminals. Terminals 7 and 31d are connected to the pickup coil in the distributor. This is an AC reluctance pickup with a resistance ranging from 450 to 1,285 ohms. The wire connected to terminal 15 carries switched ignition voltage to the module. The wire connected to terminal 31 is a ground. There are two terminal 16s (one of them may be lettered TD). One of the terminal 16s (the one that may be labeled TD), goes to the tachometer, a diagnostic connector, or a No. 1 cylinder trigger sensor. The other terminal 16 provides a ground for the negative terminal of the coil. This is the wire that fires the coil.

This system employs standard centrifugal and vacuum advance and retard.

Troubleshooting
No Start

Should the engine fail to start, connect a test light to the negative terminal of the coil. Crank the engine. As the engine is cranked, the test light should blink on and off. If the light does not blink, disconnect the connector from the ignition module. Connect an AC voltmeter between terminals 7 and 31d. Crank the engine. There should be a small AC voltage generated, about 0.5 volt or more. If there is no voltage, check the continuity in the wires between 7, 31d, and the pickup coil in the distributor. If the wires are in good condition, replace the pickup coil.

If there is an AC signal between terminals 7 and 31d, check for cranking voltage at terminal 15 of the module while the engine is being cranked. If there is no voltage at terminal 15, repair the wire from the ignition switch to terminal 15, or the wire from the battery to the ignition switch. Check continuity from terminal 31 to ground. If all the voltages are correct and there is a good ground, then check the resistance of the coil. Resistance in the primary should be between 0.33 and 1.0 ohms. (Refer to the chart in the appendix for proper resistance per specific application.) If all voltages and coil resistance are correct, replace the ignition module.

Starts But Does Not Continue to Run When the Key Is Released

This system uses a ballast resistor to limit current through the primary while the engine is running. When the engine is being cranked, battery voltage is reduced by the load of the starter. To maximize the output of the ignition system for startup, this ballast resistor is bypassed when the engine is

In most cases, Bosch electronic ignition systems are simple. In this system, terminals 7 and 31d receive the AC signals from the pickup coil in the distributor. Terminal 15 is the power supply terminal from the ignition module. Terminal 31 is the ground for the ignition module and terminal 16 is the control wire for the negative terminal of the ignition coil. An auxiliary terminal 16, or TD, can provide the signal to operate an instrument panel tachometer.

cranked. If the engine dies when the key is released, the most likely culprit is an open in the ballast resistor. Another possibility is a defective ignition switch.

Dies While Driving Down the Road

Intermittent problems are the greatest nemesis of the modern automotive technician. Loose connections on any of the wires attached to the pickup coil, ignition module, or ignition coil can cause this symptom.

If all of the wires mentioned above are in good condition, disconnect the terminal block on the ignition module and connect an ohmmeter between terminals 7 and 31. The resistance should be between 450 and 1,285 ohms. If the resistance is correct, tap on the distributor, apply a vacuum to the vacuum advance, and heat the pickup with a drop light. If the resistance fluctuates, replace the pickup coil. If the resistance does not fluctuate, replace the ignition module.

Note: If the resistance of the pickup coil does not fluctuate, it neither confirms that it is good nor that the ignition module is bad. This test merely indicates which is the most likely.

Misfire at Idle

This is almost always never the fault of the Bosch Electronic Ignition system; it is almost always the fault of some component in the secondary side of the ignition system.

Before troubleshooting any misfire, it is essential to verify that the engine is in good condition. A compression test is a good starting point. If the valves are adjustable, be sure that they are properly adjusted.

With a pair of "sissy" pliers, remove and replace one plug wire at a time from the spark plugs. As each plug wire is removed, the engine rpm should drop. If one of the cylinders fails to produce as great a drop in

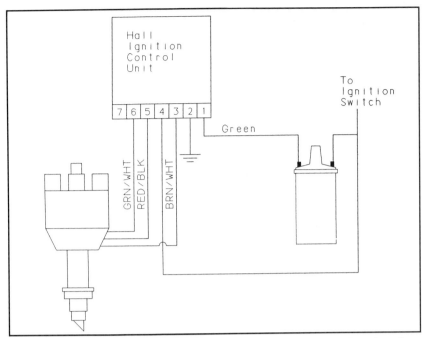

The Bosch Hall Effects ignition system is even simpler than the standard Bosch electronic ignition system. Terminal 4 receives switched ignition voltage. Terminal 1 is the control wire for the ignition coil. The other three terminals should have power for the Hall Effects, ground for the Hall Effects, and the signal from the Hall Effects to the ignition module.

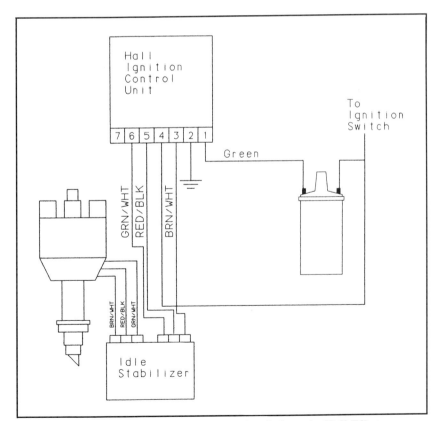

Some models equipped with the Bosch Hall Effects system also use the signals from the Hall Effect sensor as an input to the idle stabilizer.

rpm as the others, that cylinder is the source of the misfire.

Assuming the cylinder is in good condition and the valves are properly adjusted, remove the spark plug wire for that cylinder and check the resistance. The resistance should be less than 10,000 ohms per volt. If the resistance is correct, replace the spark plug. Unless the spark plugs are very new, replace them all.

Misfire Under a Load

Assuming the engine is in good condition, begin troubleshooting this problem by checking the spark plug gap. If they are gapped properly, replace the spark plugs. Even new spark plugs can misfire under a load.

If replacing the spark plugs does not solve the problem, remove the distributor cap. Inspect the wiring to the points. Frayed wiring can cause an intermittent open circuit as the vacuum advance moves the breaker plate. The intermittent open can cause a misfire.

Lack of Power

There are many things that can cause a lack of power, some related to the ignition system, some not. Begin checking this problem by confirming the engine is in good condition as are the air and fuel filters.

If a lack of power is the result of problems in the ignition system, it is likely the problem is in the timing control system. To test the timing control system, connect a timing light to the engine. Disconnect the vacuum advance and plug in the hose. With the engine at idle speed, check the timing. Now raise the engine speed to 2000 to 2500rpm. If the timing does not advance, the centrifugal advance system is not working. Inspect the distributor weights. If they are free and move easily, replace the weight springs. If the springs are weak, they will allow the timing to advance all the way prematurely, even at idle. If the weights are frozen, use penetrating oil or whatever is necessary to free them. If they are badly corroded, it may be necessary to replace the distributor.

If, or when, the centrifugal advance is working properly, with the engine still at 2000 to 2500rpm, reconnect the vacuum hose to the vacuum advance. When the vacuum hose is reconnected, the timing should advance several degrees.

Section 2

This section pertains to Bosch Hall Effects for: 1980-1986 Audi; 1981-1982 Porsche 924; 1982-1986 Saab; and 1979-1986 Volkswagen.

The Ignition Module

The ignition module has a seven pin connector. Connected to pin 1 is a green wire that provides the ground for the negative terminal of the ignition coil. Pin 2 is the ground for the ignition module. Pin 4 is connected to switched ignition voltage. The Hall Effects is connected to the ignition module through terminals 3, 5, and 6. Pin number 7 is not used. The Hall Effects device is located in the distributor.

Timing Control

This system employs standard centrifugal and vacuum advance and retard.

Troubleshooting
No Start: Basic Tests

One of the most annoying problems that can occur in an electronic ignition system is a no start. This system is among the easiest to troubleshoot, however. Begin by inserting a screwdriver in the end of a spark plug wire. Hold the screwdriver a quarter inch from ground and crank the engine. If there is a spark, then the no-start problem is not related to the ignition system. If there is no spark, place the screwdriver in the end of the coil wire and crank the engine. If there is a spark at the end of the coil wire but there is no spark at the end of the plug wire, replace the distributor cap and rotor. If the was no spark at the end of the coil wire, connect a test light to the negative terminal of the ignition coil. Crank the engine. If the test light blinks, replace the ignition coil. If the test light does not blink, confirm that there are 12 volts at terminal 4 of the ignition module and that terminal 2 is connected to a good ground. Repair as necessary.

If there are 12 volts at terminal 4 and a good ground on terminal 4, connect a tachometer to the green/white wire (terminal 6) and crank the engine. If there is a tach reading, the Hall Effects is in good condition and the ignition module should be replaced. If there is no tach signal, check for voltage on the red/black wire where it enters the distributor and continuity to ground on the brown/white wire at the distributor. Repair the wires as necessary. If the voltage to the distributor is good and there is a good ground, replace the Hall Effects pickup in the distributor.

Starts But Does Not Continue to Run When the Key Is Released

Since this system does not use a ballast resistor, it is unlikely that this symptom is caused by the Hall Effects ignition system.

Dies While Driving Down the Road

Diagnosing this symptom can be frustrating. Short of monitoring voltage and signals while driving the car, the only practical way to troubleshoot this problem is by means of an educated guess. Inspect all the wires to and from the ignition module and distributor. If they are in good condition, replace the Hall Effects. If the problem persists, replace the ignition module.

Misfire at Idle

This is almost always never the fault of the Bosch Hall Effects system; it is almost always the fault of some component in the secondary side of the ignition system.

Before troubleshooting any misfire, it is essential to verify that the engine is in good condition. A compression test is a good starting point. If the valves are adjustable, be sure that they are properly adjusted.

With a pair of "sissy" pliers, remove and replace one plug wire at a time from the spark plugs. As each plug wire is removed, the engine rpm should drop. If one of the cylinders fails to produce as great a drop in rpm as the others, that cylinder is the source of the misfire.

Assuming the cylinder is in good condition and the valves are properly adjusted, remove the spark plug wire for that cylinder and check the resistance. The resistance should be less than 10,000 ohms per volt. If the resistance is correct, replace the spark plug. Unless the spark plugs are very new, replace them all.

Misfire Under a Load

Assuming the engine is in good condition, begin troubleshooting this problem by checking the spark plug gap. If they are gapped properly, replace the spark plugs. Even new spark plugs can misfire under a load.

If replacing the spark plugs does not solve the problem, remove the distributor cap. Inspect the wiring to the points. Frayed wiring can cause an intermittent open circuit as the vacuum advance moves the breaker plate. The intermittent open can cause a misfire.

Lack of Power

There are many things that can cause a lack of power, some related to the ignition system, some not. Begin checking this problem by confirming the engine is in good condition as are the air and fuel filters.

If a lack of power is the result of problems in the ignition system, it is likely the problem is in the timing control system. To test the timing control system, connect a timing light to the engine. Disconnect the vacuum advance and plug in the hose. With the engine at idle speed, check the timing. Now raise the engine speed to 2000 to 2500rpm. If the timing does not advance, the centrifugal advance system is not working. Inspect the distributor weights. If they are free and move easily, replace the weight springs. If the springs are weak, they will allow the timing to advance all the way prematurely, even at idle. If the weights are frozen, use penetrating oil or whatever is necessary to free them. If they are badly corroded, it may be necessary to replace the distributor.

If, or when, the centrifugal advance is working properly, with the engine still at 2000 to 2500rpm, reconnect the vacuum hose to the vacuum advance. When the vacuum hose is reconnected, the timing should advance several degrees.

Asian Electronic Ignition Systems 17

Hitachi 1977-1978 Datsun Ignition

This system was used on the Federal 1977 Datsun 280ZX, Datsun 810, and California versions of the 200SX, B210, F10, 710, and pickups. In 1978, the system dropped the ballast resistor that limited primary current flow on the 1977 models and was used on all Datsun applications. The ignition module has five wires connected to it. In the 1977 version, the wires are connected to the module by screws, while on the 1978 version the wires connect to the module via a harness connector. On both versions, the wire colors are the same. The black/white wire carries switched ignition voltage to the module, while the black wire is connected to ground. The red and green wires are the leads to the pickup coil in the distributor. The blue wire provides a ground path for the coil through the ignition module.

Timing Control

This system employs standard centrifugal and vacuum advance and retard.

Troubleshooting
No Start: Basic Tests

Should the engine fail to start, insert a screwdriver in the end of a spark plug wire. Hold the screwdriver a quarter inch from ground and crank the engine. If there is a spark, the no-start problem is not related to the ignition system. If there is no spark, place the screwdriver in the end of the coil wire and crank the engine. If there is a spark at the end of the coil wire but there is no spark at the end of the plug wire, replace the distributor cap and rotor. If there is no spark at the end of the coil wire, connect a test light to the negative terminal of the ignition coil. Crank the engine. If the test light blinks, replace the ignition coil. If the test light does not blink, confirm that there are 12 volts at the black/white wire of the ignition module. If there is voltage to the black/white wire, confirm that there is continuity between the black wire at the ignition module and ground.

If there is a good voltage supply and ground for the ignition module, disconnect the red and green wires from the ignition module. Connect an AC voltmeter across the terminals and crank the engine. As the engine cranks, there should be a small AC voltage on these wires. If there is an AC voltage, check the continuity of the green wire between the module and the ignition coil.

Replace components as necessary. If all the above tests do not identify a defective component or wire, replace the ignition module.

Starts But Does Not Continue to Run When the Key Is Released

The most likely cause of this condition is a defective ballast resistor.

Dies While Driving Down the Road

Connect an ohmmeter to the

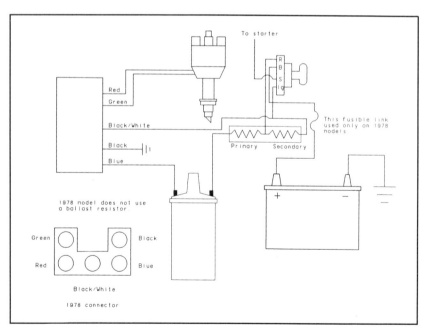

The early Hitachi electronic ignition system has five terminals on the module. An AC voltage should be detected between the green and red wires when the engine is cranked. The black/white wire should carry switched ignition voltage, the black wire is the ground. The blue wire is the control wire to the negative terminal of the ignition coil. (Note that the 1977 version had all the wires connected to the module with screws; the 1978 model used the connector illustrated in the lower right corner.)

blue and red wires. The resistance should be about 720 ohms. Heat the pickup coil with a drop light, tap on the pickup coil. If the resistance varies, replace the pickup. If the resistance remains stable, replace the ignition module. This type of troubleshooting is not accurate and often leads to replacing the wrong component; however, if the ignition system does not fail while you are testing it, there is no alternative.

Misfire at Idle

This is almost always never the fault of the electronic system; it is almost always the fault of some component in the secondary side of the ignition system.

Before troubleshooting any misfire, it is essential to verify that the engine is in good condition. A compression test is a good starting point. If the valves are adjustable, be sure that they are properly adjusted.

With a pair of "sissy" pliers, remove and replace one plug wire at a time from the spark plugs. As each plug wire is removed, the engine rpm should

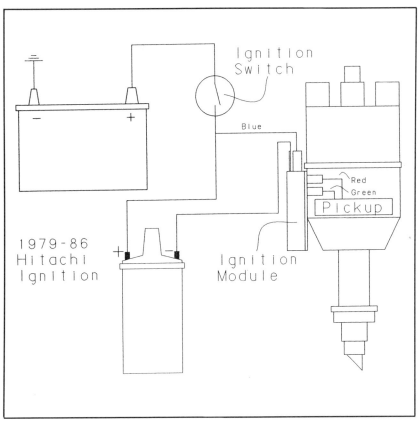

The 1979 and later Hitachi models put the ignition module on the side of the distributor. The blue external wire supplies switched ignition voltage to the module. The other external wire connects to the negative terminal of the ignition coil. The red and green wires connect to the pickup coil inside the distributor.

This is a late-model Hitachi distributor.

Many Hitachi distributor rotors have a set screw. Trying to remove the rotor without first removing the screw can prove to be frustrating and destructive.

drop. If one of the cylinders fails to produce as great a drop in rpm as the others, that cylinder is the source of the misfire.

Assuming the cylinder is in good condition and the valves are properly adjusted, remove the spark plug wire for that cylinder and check the resistance. The resistance should be less than 10,000 ohms per volt. If the resistance is correct, replace the spark plug. Unless the spark plugs are very new, replace them all.

Misfire Under a Load

Assuming the engine is in good condition, begin troubleshooting this problem by checking the spark plug gap. If they are gapped properly, replace the spark plugs. Even new spark plugs can misfire under a load.

If replacing the spark plugs does not solve the problem, remove the distributor cap. Inspect the wiring to the points. Frayed wiring can cause an intermittent open circuit as the vacuum advance moves the breaker plate. The intermittent open can cause a misfire.

Lack of Power

There are many things that can cause a lack of power, some related to the ignition system, some not. Begin checking this problem by confirming the engine is in good condition as are the air and fuel filters.

If a lack of power is the result of problems in the ignition system, it is likely the problem is in the timing control system. To test the timing control system, connect a timing light to the engine. Disconnect the vacuum advance and plug in the hose. With the engine at idle speed, check the timing. Now raise the engine speed to 2000 to 2500rpm. If the timing does not advance, the centrifugal advance system is not working. Inspect the distributor weights. If they are free and move easily, replace the weight

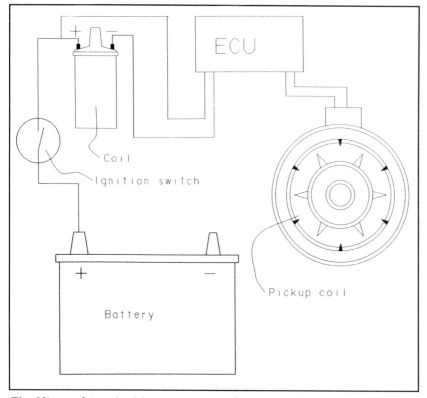

The Nippondenso ignition system uses a standard AC pickup coil. One of the other two wires is connected to the positive terminal of the coil to supply power to the ignitor, and the other is connected to the negative terminal of the coil to fire the coil.

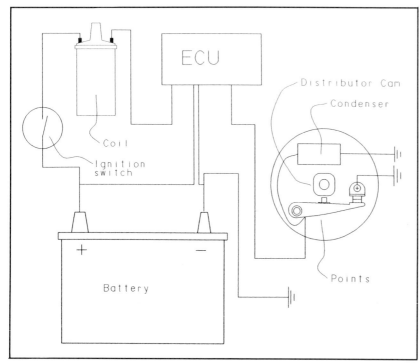

The old 20R Toyota engine used a set of low current flow points instead of a pickup coil to signal the ignition module when to fire the coil.

springs. If the springs are weak, they will allow the timing to advance all the way prematurely, even at idle. If the weights are frozen, use penetrating oil or whatever is necessary to free them. If they are badly corroded, it may be necessary to replace the distributor.

If, or when, the centrifugal advance is working properly, with the engine still at 2000 to 2500rpm, reconnect the vacuum hose to the vacuum advance. When the vacuum hose is reconnected, the timing should advance several degrees.

Hitachi 1979 Datsun/Nissan Ignition

This system is an updated version of the Hitachi electronic ignition system that was used in 1977 and 1978. The ignition module mounts on or in the distributor where it receives an engine speed reference signal from an AC pickup coil through a pair of wires. A third wire supplies switched ignition voltage, and a fourth wire is connected to the negative terminal of the ignition coil. The ground for the ignition module is through its mounting. A standard centrifugal/vacuum advance system is employed to modify ignition timing as the engine is running.

Troubleshooting
No Start: Basic Tests

Begin by using the screwdriver test in the coil wire and a spark plug wire to confirm that the problem is not in the secondary part of the ignition system. (See procedure previously mentioned in this chapter.) Crank the engine with a test light connected between the negative terminal of the ignition coil and ground. If the test light blinks, replace the ignition coil. If the test light does not blink, check for power to the ignition module. You should be able to measure about 12 volts between the positive terminal of the ignition module and the body of the distributor when the key is on. Check for an AC voltage from the pickup coil while the engine is cranking. If it is approximately 0.5 volt or more, the pickup coil is good. Check the continuity of the wire between the distributor (ignition module) and the ignition coil. If there is continuity, replace the ignition module.

Honda/Acura 1979-1986

Honda adopted the Hitachi electronic ignition system for many of its applications in 1979. When Honda introduced the Acura luxury line in 1986, it was also equipped with the Hitachi system.

The ignition module, referred to by Honda as an ignitor, has five wires connected to it. The black wire is the ground for the ignition. The red and blue wires are connected to the pickup coil in the distributor. A second blue wire connects the ignitor to the ignition coil negative terminal. A black/yellow wire connects the ignitor to the positive terminal of the ignition coil. This wire supplies power to the ignitor.

Wire Data
Black
Continuity to ground

The Toyodenso ignitor is located on the rear engine compartment bulkhead or on a fender well.

Black/Yellow
12 volts when the key is on
Between Red and Blue to distributor
AC voltage when the engine is cranked
Blue to ignition coil
Test light should blink when engine is cranked

Subaru 1978-1986

There are five wires connected to the electronic ignition module. Two of the wires run to the pickup coil, which sends an AC signal to the ignition control unit. One wire goes to coil negative while a fourth wire carries power to the module. The last

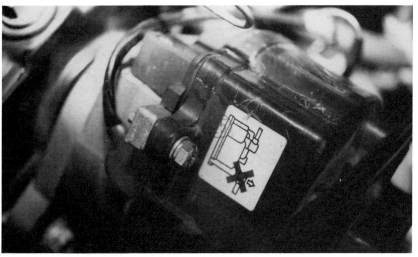

The coil on the Toyota Tercel is located in the distributor cap, a feature it shares with the GM HEI ignition system. When connected to an engine analyzer, a special adapter is required to get secondary ignition patterns.

The Toyota Tercel distributor mounts directly on, and is driven by, the camshaft.

When the negative terminal of the coil on Toyota applications is difficult to get to, a tach signal can be picked up in the diagnostic terminal. Connect the tachometer to terminal Tc.

Some Toyota applications are equipped with an altitude pressure sensor. There are three wires connected to this sensor. One wire should have 5 volts, one should be a ground with 0 volts, the last should have a voltage that varies as the pressure altitude changes.

wire is ground for the module.

Note that after 1981, the ignition control unit is located in the distributor. Since the case of the ignition control unit provides its ground, there are only four wires, 12 volts power, ignition coil negative, and two wires to the pickup coil.

Subaru Turbo Models 1983

These applications have four terminals on the ignition ignitor. Additionally, there is a Knock Control Unit (KCU) that has seven wires connected to it. A black/yellow wire connects the phase control signal terminal of the KCU. This wire carries a signal that causes the timing to retard if the knock sensor detects a knock. A black/white wire carries switched ignition voltage to the ignition coil, ignitor, and KCU. A black/red wire links the ignitor to the ignition timing terminal of the KCU. This wire carries the timing reference signal generated by the ignitor's built-in pickup to the KCU. A yellow wire connects from the ignitor to ignition coil negative. This is the wire that grounds and fires the coil. A green/red wire connects the knock sensor to the KCU. A third terminal running between the knock sensor and the KCU provides a shield against radio frequency and electromagnetic interference. The last wire, black, is connected to the KCU and provides ground.

Wire Data

Black/Yellow

Continuity between ignitor and KCU

Black/White

12 volts with ignition on

Brown/Red

Pulse detected on tach/dwell meter when engine cranked or running

Green and Red

Continuity with high resistance at the KCU terminal

Shield

Open to Green wire and Red wire

Black

Continuity to ground

Nippondenso/Toyodenso

The Nippondenso and Toyodenso systems are similar to other Asian systems. Typically, a pickup coil sends an AC sine wave along two wires to the ignitor. The ignitor receives voltage from the ignition switch and is connected to ground either by a wire or through its case.

Aftermarket Ignitions 18

Modern engines have electronic and computer-controlled timing systems. The dwell and timing are at the will of the onboard computer. For this reason most cars that are destined for street use will get little or no benefit from expensive ignition modifications. In fact, stock ignition systems of the eighties and nineties rival and exceed the abilities of racing ignition systems of earlier decades. Many stock ignition systems have a theoretical maximum ignition output voltage in excess of 90,000 volts. Additionally, power can be up to 50 or 60 watts across the tip of the plug.

In many places around the country, replacement of stock ignition systems may be either illegal or subject to strict guidelines. If you plan to replace any part of the ignition system with a nonstock unit be sure to check its legality with your local automotive emissions agency.

However, older engines with point/condenser ignition systems can benefit greatly from the installation of some of the new electronic ignition systems.

Replacement Distributors

Replacement distributors are a popular way to jazz up the ignition system during a performance modification. Back in the old days (sixties and seventies) there was a lot to be gained with performance distributors. Dual point distributors offered improved coil saturation (longer dwell) with little risk of the points arcing. With a single point distributor the only way to increase the dwell was to decrease the point gap. While this could efficiently increase coil saturation and improve the quality of the spark, it had the significant disadvantage of never allowing the contact of the points to get far enough apart to prevent arcing. The result was arcing points and really lousy performance from the ignition system.

In a classic, performance-designed dual point distributor, the point rubbing blocks were staggered with respect to the distributor cam. One set of points would close, completing the current path for the ignition coil. A short time later (say ten degrees later) the second set of points would close. If the engine in question were an eight-cylinder, about 30 degrees of crank rotation after

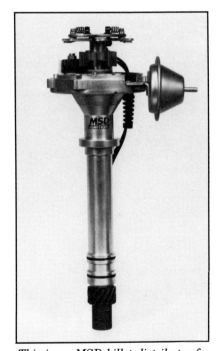

This is an MSD billet distributor for use on a Chevrolet V-8 engine. Its vacuum advance (mounted in the cannister near the top on the right in this photo) enhances timing at partial throttle to improve performance or fuel economy, sometimes both.

This MSD ignition delivers what its name states—multiple spark discharge—which allows the spark plug to fire several times during the power stroke. This particular model also features boost control, which will safeguard against detonation.

the first set of points closed, they would open. About ten degrees or so later the second set of points would open and the current path for the ignition coil would be broken. The magnetic field created by the primary would collapse and a high voltage would be created in the secondary. The benefit of a dual point distributor was to allow long point dwell.

Today the primitive technology of the dual point distributor can be completely eclipsed by electronic ignition systems. Coil saturation can be controlled to extremely long dwells with no risk of point arc. In fact many systems detect the collapse of the spark across the gap of the spark plug and begin coil saturation immediately after that collapse. In these systems the coil will always be saturated to the maximum possible. The only way it could be longer is by killing the spark prematurely.

Electronic Retrofit

Many years ago I worked for a Mercedes dealership in Fort Worth, Texas. As a service to the customers we always recommended during a regular tune-up that the standard point/condenser ignition system be replaced with the Mallory Unilite Breakerless Electronic Conversion ignition system. They were easy to install, and I did dozens of the conversions with little or no problem through the mid-eighties. As factory point/condenser applications became virtually non-existent in the late eighties, this installation has become less common. Still, that particular Mallory ignition and other similar products remain good choices over simply installing new points.

Not only Mallory, but Accell, MSD, and others market retrofit electronic ignition systems. Not all of the retrofit systems use optical triggers; some have reluctance pickups and some have Hall Effects pickups. Almost all install with basic hand tools and something to drill holes to mount the module.

Replacement Control Modules

In a way, this is a tough subject to discuss. Replacement ignition modules range in type from simple factory clones to units with timing advance and retard systems. The more sophisticated timing advance/retard modules are a good choice when there have been modifications to the engine or fuel system. An example is a case where the vehicle has been converted to run on both gasoline and natural gas.

Natural gas and M85 (methanol and gasoline blends) have a much higher octane than gasoline. This means that combustion and the propagation of the flame front is slower. This dramatically reduces the tendency of the engine to detonate. This allows the timing to be advanced more during normal operation of the engine on the alternative fuel. These slower burning alternative fuels also have a much lower heat (power) content than does gasoline. This demands that the timing be advanced more. MSD Ignition and its sister company, Autotronics, make a wide range of timing-adjustable ignition modules for just such a purpose. Some of these switch automatically when the driver switches to the alternative fuel, while others are controlled from the driver's seat.

Adjustable Timing Control

The first "racing" engine I ever built was an anthology of discarded and junk-yard parts. The engine ended up having an unbelievably high compression ratio. To get the engine started it was necessary to severely retard the timing. Once the engine was running the retarded timing severely limited engine performance. Readjusting the timing every time the engine was started proved to be inconvenient. The solution was a choke cable mounted under the dash. When the choke cable was pushed in it would rotate the breaker plate to a retarded position to get the engine started. After the engine was started it would be pulled out to advance the timing to where the driver felt he was getting the best performance. Today, this is done with electronics.

What if you just want more performance? Can a timing control ignition module help? Back when "2001" was *the* film to see on the big screen, a fellow would by a new Camaro or Barracuda and the first thing he would do was "bump up the timing." This advancing of the ignition timing may or may not have really helped the performance of the engine, depending on whether the "bump" was accompanied by a corresponding "bump" in the quality of the fuel.

Detonation Control

With the quality of fuel varying widely today across North America, a driver who is concerned about performance should, if it's legal in his jurisdiction, consider a timing control device. One of the first questions a mechanic working in southwest Canada—say, the Vancouver area—asks a customer who has suddenly developed a driveability problem is, "Did you just buy gas in the States?" One of the first questions a mechanic working in the southwest US—say, the San Diego area—asks a customer who has suddenly developed a driveability problem is, "Did you just buy gas in Mexico?" Does gasoline get progressively worse as you travel south? Lower quality gas has a tendency to be lower octane.

When running on the lower octane, performance is reduced and the chance of detonating is greatly increased. The adjustable timing control module allows the ignition timing to be adjusted for the fuel "back home," yet allows the driver to tune the timing while driving the car to match the needs of the fuel in any other location.

Performance vs. Emissions

For those to whom performance means a drop in the elapsed time—to travel the world a quarter-mile at a time—a timing control ignition module can also be advantageous. If local emission authorities permit the installation of such a module, it would allow the engine to be operated with an emission-legal timing on the street, then allow the driver to make adjustments to timing on the fly when at the track.

If you are more interested in getting your motor home through the Adirondack Mountains than winning a grudge match at the local drag strip, a timing control ignition module can be helpful. From a legal perspective, this is a little touchier than tuning the timing at the track. Here, the intent is to alter timing on a vehicle that has been type-certified with the factory setting with the intent of operating the vehicle on what is likely a federally funded highway. This puts you at risk of violating local emission regulations wherever you are. My only suggestion is to fill the back of that motor home with attorneys having emissions expertise in all of the jurisdictions through which you will be traveling.

Advancing the timing is not, and never really has been, an answer to improving power. Sometimes it does, sometimes it does not. Advancing the timing will only improve performance if no detonation occurs at the advanced setting.

Timing Control and Boost

Perhaps one of the most critical times to use variable timing control is when the ignition system is not equipped with a detonation control system and boost has been added to the engine. A turbocharger or supercharger creates a high pressure in the intake system. This higher pressure in the intake causes a higher pressure in the combustion chamber. Higher combustion chamber pressures mean a greater tendency to detonate.

If you have added a turbocharger or supercharger to your motor home's engine with the idea of getting it through the Adirondacks easier and faster, you may be disappointed. The high boost pressure may cause enough detonation at stock timing to negate the benefits of the higher cylinder pressures. Yet setting the timing to a retarded position can significantly affect performance and fuel economy at lower boost pressures. The answer is a dash-mounted, driver-controlled timing control.

Multiple Spark Discharge

One of the more interesting developments in ignition systems is the multiple spark systems. One of the leaders in this aftermarket technology is MSD Ignition of El Paso, Texas. Their name even says what they do: MSD stands for Multiple Spark Discharge. A multiple spark discharge system allows the spark plug to be fired several times during the power stroke. This is particularly advantageous in the rich environment of a racing engine of a stock engine under extreme load. Rev limiting modules are available to prevent over-revving of the engine.

Crank Trigger

For those who have "Hurst" permanently branded in the palm of their hand, the ultimate trick setup for ignition systems is the crank trigger. In performance engines, timing accuracy can be lost as a result of flex and slack in the timing chain, camshaft, and distributor drive. Since a crank trigger system mounts directly on the crankshaft, this flex, this loss of maximum power is eliminated.

This distributor drops in to older GM engines as a direct and easy replacement for the stock point/condenser ignition system. The system offers maximum coil saturation up to the highest speeds, variable dwell, and all of the other benefits of the ignition systems used in later model GM applications. Installation is merely a matter of replacing the

A crank trigger such as this one for a small-block Ford engine delivers maximum accuracy in timing control.

Note that the pickup has two wires coming out of it. It produces an AC signal to the module.

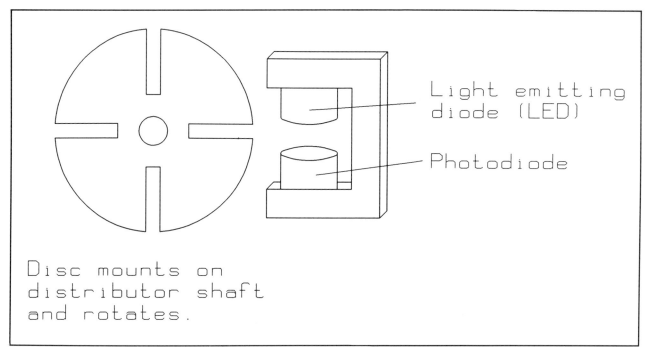

Systems such as the Mallory Unilite ignition system replace a point/condenser ignition setup with a far more reliable LED/photodiode setup. A light emitting diode (usually infrared, invisible light) sits opposite an optical receiving device such as a photodiode or phototransistor. An armature is rotated between the LED and the receiver, and as the armature rotates, light alternately falls on and is kept from falling on the receiver. The current flowing through the receiver is turned on and off creating a square wave with a frequency directly proportional to armature rotation.

stock point/condenser distributor with this unit.

Replacement of a stock point/condenser distributor with a modern electronic ignition distributor will give the performance enthusiast improved coil saturation. There are two times when maximum coil saturation is critical; at high engine speeds and when the ignition system has damaged or worn components.

The Mallory Unilite ignition system replaces a point/condenser ignition system with a far more reliable LED/photodiode setup. Eliminating the breaker points reduces the need for maintenance tune-ups.

The MSD system fires the spark plug several times during the power stroke. A typical ignition system fires the spark plug only once. The advantage is easier starting, a more complete combustion of the air/fuel mass and less of a tendency to foul the plugs. Rev limiting modules are available from MSD. These modules prevent the engine from exceeding a specified RPM and being damaged.

A crank trigger allows for maximum accuracy in timing control. Eliminating the flex and lash between the crankshaft and the distributor, these systems insure that the timing will always be perfect. Note that the pickup in the photograph has two wires coming out of it. It produces an AC signal to the module.

Ignition Specifications *Appendix*

Ignition System	Coil Primary ohms	Resistance Secondary ohms	Ballast Resistance Ohms	Pickup coil resistance Ohms	Pickup coil air gap Inches
Accura Integra 1986-	1.2-1.5	9,040-13,560	None	650-850	N/A
Accura Legend 1986-	0.35-0.43	9,040-13,560	None	650-850	N/A
AMC HEI (GM Ignition) 1980-81	0-1.0	Infinity	None	500-1,500	N/A
AMC HEI (GM Ignition) 1982-	0-1.0	6,000-30,000	None	500-1,500	N/A
AMC Reliance/Encore	0.4-0.8	2,500-5,500	None	100-200	N/A
AMC/Jeep BID 1975	1.0-2.0	8,000-12,000	None	1.6-2.4	0.05
AMC/Jeep BID 1976	1.0-2.0	9,000-15,000	None	1.6-2.4	0.05
AMC/Jeep BID 1977	1.25-1.4	9,000-15,000	None	1.6-2.4	0.05
Audi 1977-80	0.95-1.4	5500-8000	1	890-1285	0.01
BMW 1977-80	0.4	7300-8300	0.4 and 0.6	520-700	0.014-0.028
BMW 1984-	0.7-0.9	7300-8300	0.4 and 0.6	520-700	0.014-0.028
Bosch Hall Effects, all	0.52-0.76	2,400-3,500	None	N/A	N/A
Chevrolet Spectrum 1984-	1.1-1.4	12,100-14,900	None	140-180	0.008-0.016
Chevrolet Sprint 1984-	1.1-1.4	10,800-16,200	None	130-190	0.008-0.016
Chevy LUV 1981	1.13-1.53	10,000-13,800	1.1-1.4	130-190	0.008-0.016
Chevy LUV 1982	0.83-1.02	12,150-14,850	1.1-1.4	130-190	0.008-0.016
Chrysler 1972-79	1.41-1.79	8,000-12,000	0.5/1.12-1.38 dual	150-900	0.006-0.008
Chrysler 1980-	1.34-1.55	8,000-12,000	1.12-1.38	150-900	0.008/0.012 dual
Chrysler Hall Effects	1.41-1.79	8,000-11,700	None	N/A	N/A
Chrysler Hall Effects Lean Burn	1.34-1.79	8,000-11,700	See text	N/A	N/A
Chrysler Lean Burn 1977-79	1.41-1.79	8,000-12,000	1.2	150-900	0.006
Chrysler Lean Burn 1980-87	1.34-1.79	9,000-12,200	1.2	150-900	0.006
ChryslerLean Burn 1976-77	1.34-1.79	8,000-12,000	0.5-0.6/4.75-5.75 dual	150-900	0.008/0.012 dual
Datsun 1977	0.45-0.55	8,500-12,700	0.4 and 0.9	720	0.008-0.016
Datsun 1978	0.84-1.02	8,200-12,700	0.4 and0.9	720	0.008-0.016
Datsun 310 1979-82	0.84-1.02	8,200-12,700	0.4 and0.9	720	0.008-0.016
Datsun 510 1979-81	0.84-1.02	8,200-12,700	0.4 and0.9	720	0.008-0.016
Datsun/Nissan 1979- 210, 280ZX	0.84-1.02	8,200-12,700	0.4 and0.9	720	0.008-0.016
Datsun/Nissan 1979- 810 Stanza	0.84-1.02	8,200-12,700	0.4 and0.9	720	0.008-0.016
Datsun/Nissan Pick-up 1979-	0.84-1.02	8,200-12,700	0.4 and0.9	720	0.008-0.016
Fiat 1977-82	1.1-1.7	6,000-10,000	0.85-0.95	890-1285	0.011-0.019
Ford Dura-Spark I 1977-78	0.71-0.77	7,350-9,300	None	400-800	N/A
Ford Dura-Spark I 1979	0.71-0.77	7,350-9,300	None	400-1,000	N/A
Ford Dura-Spark II 1977-78	1.0-2.0	7,000-13,000	1.05-1.15	400-800	N/A
Ford Dura-Spark II 1979	1.13-1.23	7,000-13,000	1.05-1.15	400-1,000	N/A
Ford Dura-Spark II 1980	1.13-1.23	7,700-9,600	1.05-1.15	400-1,000	N/A
Ford Dura-Spark II 1981	1.0-1.2	7,000-13,000	1.05-1.15	400-1,000	N/A
Ford Dura-Spark II 1982-	0.8-1.6	7,700-10,500	0.8-1.6	400-1,000	N/A
Ford Dura-Spark III 1979-81	1.13-1.23	7,700-9,600	1.05-1.15	N/A	N/A
Ford Dura-Spark III 1982-84	0.8-1.6	7,700-10,500	0.8-1.6	N/A	N/A
Ford SSI 1974-1976	1.0-1.2	7,000-13,000	1.4	400-800	N/A
Ford TFI (EFI)	0.3-1.0	8,000-11,500	None	N/A	N/A
Ford Thick Film EFI	0.3-1.0	8,000-11,500	None	N/A	N/A
Ford Thick Film Non-EFI	0.3-1.0	8,000-11,500	None	650-1,300	N/A
GM C3I 3 coil type	N/A	Less than 15,000	None	N/A	N/A
GM C3I 3-in-1 coil pack	N/A	5,000-7,000	None	N/A	N/A
GM HEI (External Coil) 1974-79	0-1.0	6,000-30,000	None	500-1,500	N/A
GM HEI (External Coil) 1980-	0-1.0	See text	None	500-1,500	N/A
GM HEI (Internal Coil) 1974-75	0-1.0	6,000-30,000	None	500-1,500	N/A

GM HEI (Internal Coil) 1976-79	0-1.0	Infinity	None	500-1,500	N/A
GM HEI (Internal Coil) 1980-	0-1.0	See text	None	500-1,500	N/A
GM HEI 1980-81 Int. coil	0-1.0	6,000-30,000	None	500-1500	N/A
GM HEI EST 1980-	0-1.0	See text	None	500-1,500	N/A
Honda 1979 except Civic	1.78-2.08	8,800-13,200	None	600-800	N/A
Honda 1980 All others	1.78-2.08	8,800-13,200	None	800-1200	N/A
Honda Calif. Accord & Prelude 80	1.06-1.24	7,400-11,000	None	800-1200	N/A
Honda Civic 1980	1.0-1.3	7,400-11,000	None	800-1200	N/A
Honda Civic 1981-83	1.0-1.3	7,400-11,000	None	N/A	N/A
Honda Civic 1984-85	1.24-1.46	8,000-12,000	None	N/A	N/A
Honda Prelude & Accord 1981-85	1.06-1.24	7,400-11,000	None	N/A	N/A
Isuzu 1981-	1.13-1.53	10,000-13,800	1.1-1.4	130-190	0.008-0.016
Mercedes Benz 1977-79	0.33-0.46	7,000-12,000	0.4 and 0.6	450-750	0.01
Mercedes Benz 1981 6 Cylinder	0.7	8,000-10,000	None	500-700	N/A
Mercedes Benz 1982- V8	0.38-0.42	8,000-10,000	None	500-700	N/A
Mercedes Benz, other	0.38-0.46	8,000-11,000	0.4 and 0.6	500-700	N/A
Mitsubishi Electronic Ign.	0.70-0.85	9,000-11,000	None	920-1120	N/A
Nissan 1980-86 200SX, Pulsar	0.84-1.02	7,300-11,000	0.4 and0.9	720	0.008-0.016
Nissan Maxima 1982-84	0.84-1.02	8,200-12,700	0.4 and0.9	720	0.008-0.016
Nissan Sentra	0.84-1.02	7,300-11,000	0.4 and0.9	720	0.008-0.016
Peugeot 604 1977-1980	0.33-0.46	7,000-12,000	0.5 and 0.5	485-700	N/A
Peugeot with Ducellier Ignition	0.48-0.61	9,000-11,000	None	900-1100	0.012-0.02
Porsche 1977-80 924	1.0-1.35	5,500-8,000	0.4 and 0.6	890-1285	0.01
Porsche 1977-83 928	0.33-0.46	7,000-12,000	1.0-1.5	485-700	0.01
Porsche 1984-86	0.33-0.46	7,000-12,000	0.4 and 0.6	485-700	N/A
Renault except Fuego Turbo	0.48-0.61	9,000-11,000	None	N/A	0.012-0.024
Renault Fuego Turbo	0.4-0.8	2,500-5,500	None	100-200	0.02-0.06
Saab 1978-81	1.05-1.35	5,500-8,500	0.4 and 0.6	895-1285	N/A
Subaru 1977-78	0.81-0.99	8,500-12,900	0.9	600-850	0.012-0.016
Subaru 1977-78 Nippondenso	1.33-1.63	11,100-13,700	1.4	130-190	0.008-0.016
Subaru 1977-78 Nippondenso	1.13-1.38	10,800-14,600	None	130-190	0.008-0.016
Subaru 1979-80	1.17-1.42	7,800-11,600	None	600-850	0.012-0.016
Subaru 1979-80 Nippondenso	1.33-1.63	12,600-15,400	1.1-1.3	130-190	0.008-0.016
Subaru 1981-82 Nippondenso	1.06-1.30	12,150-14,850	1.4-1.6	130-190	0.008-0.016
Subaru 1981-86 Non-Turbo	1.04-1.27	7,360-11,040	None	N/A	0.012-0.020
Subaru 1983-86 Turbo	0.84-1.02	8,000-12,000	None	N/A	0.012-0.020
Suzuki Samarai	1.35-1.65	11,000-14,500	None	130-190	0.008-0.016
Toyota 1977	1.35-1.65	12,800-15,200	1.3-1.7	130-190	0.008-0.016
Toyota 1978-79	1.3-1.7	12,000-16,000	1.1-1.4	130-190	0.008-0.016
Toyota 1980 Calif. Celica, Corona	0.8-1.0	11,500-15,500	1.1-1.4	130-190	0.008-0.016
Toyota 1980 Calif. Corolla, Pick-up	0.8-1.0	11,500-15,500	1.1-1.4	130-190	0.008-0.016
Toyota 1980 Cressida, Supra	0.5-0.6	11,500-15,500	1.1-1.4	130-190	0.008-0.016
Toyota 1980 Fed. Celica, Corona	0.5-0.6	11,500-15,500	1.1-1.4	130-190	0.008-0.016
Toyota 1980 Land Cruiser & Fed. P/U	1.3-1.7	12,000-16,000	1.1-1.4	130-190	0.008-0.016
Toyota 1980 Tercel & Fed. Corolla	0.4-0.5	8,500-11,500	1.1-1.4	130-190	0.008-0.016
Toyota 1981-82 Celica, Corona	0.8-1.1	11,500-15,500	1.1-1.4	130-190	0.008-0.016
Toyota 1981-82 Corolla, Tercel	0.8-1.1	11,500-15,500	1.1-1.4	130-190	0.008-0.016
Toyota 1981-82 Cressida, Supra	0.8-1.1	11,500-15,500	1.1-1.4	130-190	0.008-0.016
Toyota 1981-82 Land Cruiser	0.8-1.1	11,500-15,500	1.1-1.4	130-190	0.008-0.016
Toyota 1981-82 Pickup	0.4-0.5	8,500-11,500	1.1-1.4	130-190	0.008-0.016
Toyota 1983 Tercel	0.4-0.5	7,700-10,400	1.1-1.4	130-190	0.008-0.016
Toyota 1983- Camry, Corolla, Van	0.3-0.5	7,500-10,500	1.1-1.4	140-180	0.008-0.016
Toyota 1983- Celica	0.8-1.1	10,700-14,500	1.1-1.4	130-190	0.008-0.016
Toyota 1983- Cressida, P/U, Supra	0.4-0.5	8,500-11,500	1.1-1.4	130-190	0.008-0.016
Toyota 1983- Land Cruiser	0.5-0.7	11,500-15,500	1.1-1.4	130-190	0.008-0.016
Toyota 1983- Starlet	1.2-1.7	10,700-14,500	1.1-1.4	140-180	0.008-0.016
Volvo 1977-79 4 Cylinder	1.0-2.0	7,000-12,000	1	850-1250	0.01
Volvo 1977-83 V6	1.0-2.0	7,000-12,000	1	540-660	0.01
Volvo 1980-83 4 Cylinder	1.0-2.0	7,000-12,000	1	950-1250	0.01
Volvo 1984-86 760 GLE	0.5-1.0	7,500-12,000	0.4-0.6	540-660	0.012
Volvo all others	1.0-2.0	7,500-12,000	1	950-1250	0.01

Index

AC voltmeter, 11, 19, 103
Adjustable timing control, 155
Aftermarket ignition components, 154
Air/fuel ratio, 24-25
ALCL terminals, GM, 139-141
Alcohol contamination, 28-29
AMC/Jeep Breakerless Inductive Discharge (BID), 42
AMC/Renault Ducellier Electronic Ignition, 53
Amplitude, 15
Amps, 11-12
ASD relay, 93

Battery, 5-7, 31
Block learn multiplier (BLM), 134
Bosch Electronic Ignition System, 144

Carb switch, 75
Centrifugal advance, 37-38, 44, 50, 60, 65, 70, 77
Chrysler EFI and Optical Ignition, 81
Chrysler Electronic Ignition, 57
Chrysler Hall Effects Ignition, 63
Circuit defects, 17
Circuits
    open, 17
    short, 17
Coil wire, 34, 43, 49, 54, 58, 64, 69, 76, 83-85, 96, 101, 121
Coil, 30, 34, 43, 49, 54, 58, 64, 69, 76, 83, 96, 101, 120
Combustion, spark ignition, 21-23
Compression, 24
Computer controlled coil ignition (C_3I), GM, 137-138
Condenser, 32
Control module, 57-58
Control modules, replacement, 155
Coolant temperature override (CTO) switch, 127
Crank triggers, 156
Current flow, 12
Cylinder inhibit, testing for, 29

Datsun Hitachi ignition 1977-1978, 148-151
Datsun/Nissan Hitachi ignition 1979, 151
Dead hole, testing for, 29
Detonation control, 155
Detonation, 128
Direct Ignition System (D.I.S.), GM, 138-139
Distributor cap, coil-in-cap, 129-130
Distributor caps and rotors, 25-27, 34-35, 43, 49, 54, 59, 64-65, 69, 76, 85, 96, 101, 121
Distributor pickup, 48-49, 95-96
Distributorless ignition system (DIS), 130-132
Distributors, aftermarket, 154
Duty cycle, 16-17

E-Core coil, 109
Electromagnet voltage sources, 7
Electromagnetism, 7-9
Electronic Control Unit (ECU), 42, 48, 63, 69, 73-74, 95, 118
Electronic Engine Control systems (EEC-II,

EEC-III, EEC-IV), Ford, 101, 110
Electronic Lean Burn 1976-1977, 73
Electronic Spark Control (ESC), Chrysler, 79-80
Electronic Spark Control (ESC), GM, 79-80, 142-143
Electrostatic discharge (ESD), 18
Engine Control Module (ECM), GM, 126-127, 130

Filters, air and fuel, 27
Ford Duraspark III, 101
Ford Solid State Ignition (SSI), 95
Frequency, 15

General Motors (GM) Ignition System, 118

Hall Effects, 10, 63, 81-83
High Energy Ignition (HEI) system, GM, 118
Honda/Acura ignition, 1979-1986, 151

Idle air control (IAC) motor, 135
Ignition control module, 53-54
Ignition diagnostic monitor (IDM), 115
Ignition switch, 30-31, 43, 49, 54, 58, 63, 69, 75, 96, 101, 120
Impedance, 10-11
Induction, 7
Inductive ammeter, 12
Integrator (INT), 133

Kirchoff's Second Law, 14

Malfunction indicator light (MIL), 112
Manifold absolute pressure (MAP) sensor, 87-88
Meters
    analog, 10
    digital, 10
    duty cycle, 16
    dwell, 16
    oxygen, 7
    pulse width, 17
Mitsubishi Electronic Ignition, 68
Multiple spark discharge, 156
Multipoint injection (Chrysler), 90-92

Nippondenso/Toyodenso ignition, 153

Ohm's Law, 13-14
Osciloscope, 11, 17, 19

Parallel robbery, 17-18
Photoelectric voltage sources, 7
Pickup coil, 18-19, 54, 58, 63, 69, 74-75, 101, 107-109, 118-120
Point/condenser ignition system, 30
Points, 32-33
Pulse width, 17

Resistance, 12-13
Resistor, ballast, 31-32
Resistors, 13
Rotors, 25

Self-test output (STO), 111

Self-test input (STI), 111
Sensor coil, 42-43
Sensors
    coolant, 75, 88-89
    Hall Effects, 19-20
    knock, 128
    Lambda, 29
    optical, 20, 87-88
    oxygen, 134-135
    Profile Ignition Pickup (PIP), 105-107, 109-110, 114-115, 117
    speed, 68
    TDC, 56
Shunt ammeter, 12
SMEC system (Chrysler) 1987-1990, 92
Solenoids, 9-10
Solid State Ignition (SST), 48
Spark plug wires, 25, 35-36, 43-44, 49, 55, 59, 65, 70, 77, 86, 96, 102, 122
Spark plugs, 23-25, 36-37, 44, 50, 55, 59-60, 65, 70, 77, 87, 97, 103, 122
SPOUT wire and connector, 109
Subaru ignition, 1978-1986, 151-153
Subaru turbo model ignition, 1983, 153

Tachometer, 66
TFI module, 107
Thermo-vacuum switch (TVS), 127
Thick Film Integrated (TFI) ignition, Ford, 105
Throttle bodies (Chrysler), 90-92
Throttle position transducer, 75
Timing control, 37, 44, 40, 55, 60, 65, 70, 77, 87-88, 97, 103, 123
Trouble codes
    Chrysler, 93-94
    Ford, 114-117
    GM ignition systems, 141-143
Troubleshooting
    AMC/Jeep BID, 45-47
    AMC/Renault Ducellier Electronic Ignition, 55-56
    Bosch electronic ignition systems, 144-147
    Chrysler EFI and Optical Ignition, 89-94
    Chrysler electronic ignition, 60-62
    Chrysler Hall Effects ignition, 66-67
    electronic lean burn, 78-80
    Ford EEC systems, 103-104
    Ford Solid State Ignition (SSI), 97-100
    Ford TFI ignition, 109-117
    GM Ignition systems, 123-143
    Mitsubishi Electronic Ignition, 70-72
    point/condenser ignition systems, 39-41
    Solid State Ignition, 50-52
Tune-up procedures, GM ignition systems, 136

Vacuum advance, 38, 44, 50, 60, 65, 70, 77, 97
Vacuum transducer, 75-76
Variable Reluctance Transducer (VRT)—also see Pickup coil, 18-19, 81
Voltage, 10

Watts, 13
Waves, sine and square, 14